Deep Learning in Object Detection and Recognition

Xiaoyue Jiang • Abdenour Hadid • Yanwei Pang
Eric Granger • Xiaoyi Feng
Editors

Deep Learning in Object Detection and Recognition

Editors
Xiaoyue Jiang
School of Electronics and Information
Northwestern Polytechnical University
Xi'an, Shaanxi, China

Yanwei Pang
School of Electrical and Information
Engineering
Tianjin University
Tianjin, Tianjin, China

Xiaoyi Feng
School of Electronics and Information
Northwestern Polytechnical University
Xi'an, Shanxi, China

Abdenour Hadid
Center for Machine Vision
and Signal Analysis
University of Oulu
Oulu, Oulu, Finland

Eric Granger
École de technologie supérieure
University of Québec
Montréal, QC, Canada

ISBN 978-981-10-5151-7 ISBN 978-981-10-5152-4 (eBook)
https://doi.org/10.1007/978-981-10-5152-4

This Springer imprint is published by the registered company Springer Nature Singapore Pte Ltd.
The registered company address is: 152 Beach Road, #21-01/04 Gateway East, Singapore 189721, Singapore

Preface

Object detection and recognition are key components for most computer vision systems and determine the performance of the many applications, such as tracking, retrieval, video surveillance, and image captioning. The performance of object detection and recognition heavily depends on the quality of the extracted features and robustness of classifiers, since the appearance of images may be influenced by many factors like lighting conditions, the pose, the reflectance of objects, and the intrinsic characteristics of cameras. To achieve robust detection and recognition, the extracted features that are used for verification must be invariant to lighting, pose, and other transformations. In classical applications, the gradient-based features, such as edges, local binary patterns (LBP), and scale-invariant feature transform (SIFT), have shown their robustness to the variations of lighting and pose. Besides robust features, the classifiers also need to be discriminative enough for those intertwined features. For traditional detection and recognition systems, the feature extraction and classification steps must work together to achieve good performance.

In recent years, deep neural networks have attracted more and more attentions and have been applied successfully in many field of image processing and computer vision. First, neural networks can learn to extract features from training dataset. Thus, we do not need to design features as the classical procedure, which is always very costive. Second, neural networks can extract a series of high-order nonlinear features from data during the training procedure. These nonlinear features are constructed through convolutional layers, which are also adaptive to the training data. Thus, it can represent the distribution of data more accurately than the traditional features. Third, deep neural networks are end-to-end systems. The input can be the original image and the output can be the desired results when the neural networks combine two basic procedures of detection or recognition into one framework. Due to the great capacity of deep neural networks, they have achieved great success in many computer vision tasks, including object detection and recognition.

In this book, we aim to provide a comprehensive overview for the current development of deep learning in object detection and recognition. With the introduction of deep neural networks from different aspects, the structure, characteristics, and performance of the deep neural networks can be understood more thoroughly. With the inspiration from different fields, we hope our readers can get some inspirations to design suitable deep neural networks for their own tasks.

In Chap. 1, the history of deep neural network is briefly reviewed. Then, the most used networks, namely, stacked autoencoders, deep belief networks, convolutional neural networks, recurrent neural networks, and generative adversarial networks are discussed, respectively. Through the introduction of basic concepts and theories of neural network, the readers can get the background knowledge for deep neural networks.

The application of deep neural networks in object detection is discussed in Chap. 2. There are two main approaches for using neural networks for object detection, namely, two-stage methods and one-stage methods. For the two-stage methods, proposals of possible objects are extracted first and then classified for detection. The two-stage methods, such as R-CNN, Fast RCNN, and Faster RCNN, have achieved the state of the art for the detection tasks, but the speed of these methods is relatively slow. For one-stage methods, such as OverFeat, YOLO, SSD, and RetinaNet, they predict the position and category of object simultaneously. Thus, the speed is much faster than that of the two-stage methods, but the accuracy is degraded. Finally, the pedestrian detection is used as an example to illustrate the advantage of some latest neural network structures for detection.

Besides detection, recognition is another essential task for image processing systems. In Chap. 3, the application of deep neural networks in face recognition is introduced. Even though face recognition has already been utilized in many commercial systems, the performance of current system tends to degrade significantly in extreme conditions, such as variant lighting and pose conditions. The algorithms that deal with the lighting and pose challenges are thoroughly reviewed in this chapter. Through the comparison with the traditional features, the deep neural networks showed advantages in recognition tasks. However, the networks used for recognition across lighting and pose should focus on the correlations between different local regions. Finally, a new neural network extracting local features is introduced in this chapter.

In Chap. 4, the application of deep neural networks in face anti-spoofing task is presented. In fact, with more and more applications of face recognition in commercial verification systems, the anti-spoofing technique becomes more and more important. The images that are used to fool verification systems are always recaptured from the original image. Thus, the anti-spoofing systems should be able to discriminate the difference between the original and recaptured images. In this chapter, a convolutional neural network is combined with the traditional feature to create a more robust feature for anti-spoofing tasks. Generally, the deep features are more discriminative than the traditional ones. However, for some specific applications, the deep features can perform better with the enhancement from traditional features. It brings the thoughts that the deep features and traditional features are not mutually exclusive but can improve each other.

For facial image analysis, kinship verification is a very useful application. In Chap. 5, the kinship recognition based on deep neural networks is discussed. In fact, people from the same family always share some facial attributes, which can be utilized to verify the family relationship. Deep learning methods are also introduced in this specific field to extract subtle features between family members. Compared with the traditional features, the deep features are more discriminative. Also, the deep neural networks provide a complete framework to solve this verification challenge.

In Chap. 6, the deep neural networks are applied to face recognition of video surveillance. For video surveillance, the challenges are mainly due to the significant difference between the reference images and the images captured in surveillance camera, where the illumination, pose, scale, occlusion, and intrinsic camera features are all different. Thus, more robust invariant features are required for video surveillance problems. Consequently, deep neural networks are widely applied to extract these desired invariant features for video surveillance scenarios. Besides the structure of convolutional layers, the loss function also determined the features that can extract in deep neural networks. In this chapter, a triplet loss function scheme is introduced to ensure that the extracted features can discriminate the positive samples from the negative ones.

Besides 2D object recognition, 3D object recognition is also required in real applications. In Chap. 7, deep learning-based 3D object recognition is introduced. For 3D object recognition, local features determine the performance of the system. In this chapter, the restricted Boltzmann machine is applied for the analysis of 3D data. Besides the convolutional neural networks, the restricted Boltzmann machine is another type of neural network structure that is used widely. In order to get more accurate local features for 3D data, a new circle convolution kernel is introduced for the feature extraction of 3D data in this chapter. With the sophisticated designed kernel, the neural network can extract more robust features from data.

In Chap. 8, the deep learning-based image retrieval problems are discussed. For image retrieval, it always requires efficient descriptions for the characteristics of the whole images, but not only focuses on some specific targets. Then in this application, the task is to construct compact features for the whole image from deep neural networks. Also, the scale and rotation invariant characteristics are required for the extracted deep features. A new regularization method and a pooling method are introduced in this chapter to achieve those two characteristics for deep features.

From all these eight chapters, deep learning-based methods showed great breakthroughs in the application of object detection and recognition. Researchers do adaptations of deep networks in different aspects to make them fit the specific requirement of data analysis, which include convolutional kernels, loss functions, pool scheme, regularization scheme, and the structure of the networks. We hope our readers can get some thoughts about how to apply deep neural networks in their applications and also get a taste of how to improve the deep neural networks to be more powerful for their own tasks.

Last but not least, we would like to thank the publisher for providing us with the opportunity to write this book. Also, we would like to thank all the authors and reviewers who contribute their precious thoughts for this book.

Xi'an China — Xiaoyue Jiang
Oulu, Finland — Abdenour Hadid
Montreal, Canada — Eric Granger
Tianjin, China — Yanwei Pang
Xi'an, China — Xiaoyi Feng
March 2018

Contents

Acronyms

2D	Two Dimensional
3D	Three Dimensional
AEs	Autoencoders
ANNs	Artificial Neural Networks
AP	Average Precision
ASM	Active Shape Model
BP	Back Propagation
BPBC	Bilinear Projection-Based Binary Codes
BPTT	Back Propagation Through Time
BSIF	Binarized Statistical Image Features
BoW	Bag of Word
CASIA	Institute of Automation, Chinese Academy of Sciences
CASIA-FA	CASIA Face Anti-spoofing Database
CCA	Canonical Correlation Analysis
CCDBN	Circle Convolutional Deep Belief Network
CCF	Convolutional Channel Features
CCM	Cross-Correlation Matching
CCM-CNN	Cross-Correlation Matching Convolutional Neural Network
CCRBM	Circle Convolutional Restricted Boltzmann Machine
CD	Contrastive Divergence
CDBN	Convolutional Deep Belief Network
CFN	Channel Feature Network
CFR-CNN	Canonical Face Representation Convolutional Neural Network
CIFAR	Canadian Institute for Advanced Research
CNN	Convolutional Neural Network
CNN-RBMH	Convolutional Neural Networks Restricted Boltzmann Machine Regularization Scheme for Hashing
COCO	Common Objects in Context (Microsoft dataset)
CRBM	Convolutional Restricted Boltzmann Machines
CUHK	Chinese University of Hong Kong
CoALBP	Co-occurrence of Adjacent Local Binary Patterns

DARPA	Defense Advanced Research Projects Agency
DBN	Deep Belief Network
DCCA	Deep Canonical Correlation Analysis
DCNN	Deep Convolutional Neural Network
DMML	Discriminative Multimetric Learning
DeepID	Deep Hidden IDentity Features
DeepID2	Deep IDentification-Verification features
DoG	Difference of Gaussian
EER	Equal Error Rate
EKF	Extended Kalman Filter
ERG	Extended Reeb Graphs
FC	Fully Connected
FCF	Filtered Channel Features
FDDB	Face Detection Dataset and Benchmark
FGS	Farthest Geodesic Sampling
FPLBP	Four-Patch Local Binary Patterns
FPN	Feature Pyramid Network
FR	Face Recognition
FTM	Fourier Transform Modulus
FV	Fisher Vector
FV-RBMH	Fisher Vector Restricted Boltzmann Machine Regularization Scheme for Hashing
GAN	Generative Adversarial Network
GCC	Geometric Consistency Checks
GIST	Gastrointestinal Stromal Tumors
GMLDA	Generalized Multiview Linear Discriminative Analysis
GMM	Gaussian Mixture Model
HE	Histogram Equalization
HKS	Heat Kernel Signature
HMM	Hidden Markov Models
HOG	Histograms of Gradient
HSV	Hue Saturation Value
HTER	Half Total Error Rate
HaarNet	Haar-Like Deep Convolutional Neural Networks
ICB	International Conference on Biometric
ICDAR	International Conference on Document Analysis and Recognition
ICF	Integral Channel Features
ILSVRC	ImageNet Large-Scale Visual Recognition Challenge
IR	Information Retrieval
ISI	Intrinsic Spin Image
ITQ	Iterative Quantization
KPCA	Kernel Principle Component Analysis
LBP	Local Binary Pattern
LBP-TOP	Local Binary Patterns on Three Orthogonal Planes
LCD	Liquid Crystal Display

LDA	Linear Discriminative Analysis
LFD	Light Field Descriptor
LFHN	Local Feature Hierarchy Networks
LFW	Labeled Faces in The Wild
LGD	The Longest Geodesic Distance of That Shape
LPQ	Local Phase Quantization
LR	Logistic Regression
LSH	Locality-Sensitive Hashing
LSTM	Long Short-Term Memory
MCCA	Multiview Canonical Correlation Analysis
MDR-TL	Mean Distance Regularized Triplet Loss
MIR-FLICKR	Multimedia Information Retrieval Flickr
MLPs	Multilayer Perceptrons
MNRML	Neighborhood Repulsed Metric Learning
MOP-CNN	Multi-scale Orderless Pooling Convolutional Neural Networks
MPEG-CDVS	Moving Picture Experts Group Compact Descriptors for Visual Search
MR	Miss Rate
MRF	Markov Random Field
MS-CNN	Multi-scale Deep Convolutional Neural Network
MSE	Mean Squared Error
MvDA	Multiview Discriminant Analysis
NDCG	Normalized Discounted Cumulated Gain
NF	Neighboring Features
NIP	Nested Invariance Pooling
NLP	Natural Language Processing
NNF	Non-neighboring Features
NNNF	Non Neighboring and Neighboring Features
PAM	Pose-Aware Convolutional Neural Network Models
PBPR	Patch-Based Partial Recognition
PCA	Principle Component Analysis
PDD	Projection Distance Distribution
PQ	Product Quantization
PR	Precision and Recall
PSROI	Position-Sensitive ROI Pooling
R-CNN	Region Convolutional Neural Networks
R-MAC	Regional Maximum Activation of Convolutions
RAM	Random Access Memory
RBM	Restricted Boltzmann Machines
RBMH	Restricted Boltzmann Machine Regularization Scheme for Hashing
RELU	Rectified Linear Unit
RF	Random Forests
RFG	Reference Face Graph
RGB	Red Green Red

RGB-D	Red Green Blue Depth
RNN	Recursive Neural Networks
ROC	Receiver Operating Characteristic Curve
ROI	Region of Interest
RTRL	Real-Time Recurrent Learning
ReLu	Rectified Linear Units
SAEs	Stacked Autoencoders
SDP	Scale-Dependent Pooling
SGD	Stochastic Gradient Method
SH	Spectral Hashing
SHD	Spherical Harmonic Descriptor
SHREC	Shape Retrieval Contest Datasets
SI	Spin Image
SIDF	Side-Inner Difference Features
SIFT	Scale-Invariant Feature Transform
SIHKS	Scale-Invariant Heat Kernel Signature
SKLSH	Shift-Invariant Kernels Locality Sensitive Hashing
SMCNN	Similarity Metric Based Convolutional Neural Networks
SPLE	Spatial Pyramid Learning
SPP	Spatial Pyramid Pooling Layer
SRBM	Stack Restricted Boltzmann Machines
SRBMH	Stack Multiple Restricted Boltzmann Machine Regularization Scheme for Hashing
SSF	Symmetrical Similarity Features
SSPP	Single Sample Per Person
SSR	Signed Square Rooting
SURF	Speeded Up Robust Feature
SVM	Support Vector Machine
TBE-CNN	Trunk-Branch Ensemble Convolutional Neural Network
TOP	Three Orthogonal Planes
TPLBP	Three-Patch Local Binary Patterns
VLAD	Vector of Locally Aggregated Descriptors
VS	Video Surveillance
WKS	Wave Kernel Signature
WLD	Weber Local Descriptor
YCbCr	Luminance, Chrominance Blue, Chrominance Red
mAP	Mean Average Precisions

An Overview of Deep Learning

Zhaoqiang Xia

Abstract In the last decade, deep learning has attracted much attention and becomes a dominant technology in artificial intelligence community. This chapter reviews the concepts, methods, and latest applications of deep learning. Firstly, the basic concepts and developing history of deep learning are revisited briefly. Then, five basic types of deep learning methods, i.e., stacked autoencoders, deep belief networks, convolutional neural networks, recurrent neural networks, and generative adversarial networks, are introduced according to applications of deep learning in other domains that are briefly illustrated based on the types of data, such as acoustic data, image data, and textual data. Finally, several issues facing by deep learning are discussed to conclude the trends.

1 Brief Introduction

Deep learning approaches are a class of machine learning algorithms that use many layers of nonlinear processing units for representations and transformations. Each layer uses the output from the previous layer as input and the hierarchical representations can be obtained by different levels of abstraction. These algorithms may be trained in a supervised or an unsupervised way, and their applications include pattern analysis (unsupervised) and classification (supervised). Currently, the rapid development of deep learning has been accelerated by three key reasons, i.e., massive data, powerful computation ability, and novel algorithms. In the future, more and more domains will be facilitated by deep learning and reversely make contributions back to deep learning technologies. The years ahead are full of challenges and opportunities to improve deep learning technologies and bring it to new frontiers.

Z. Xia (✉)
School of Electronics and Information, Northwestern Polytechnical University, Xi'an, China
e-mail: zxia@nwpu.edu.cn

X. Jiang et al. (eds.), *Deep Learning in Object Detection and Recognition*,
https://doi.org/10.1007/978-981-10-5152-4_1

Although the deep learning is a new technology emerging in recent years, it has a long and rich history in the last 50 years. Deep learning dates back to the 1940s and rises from the artificial neural networks (ANNs) in the fields of artificial intelligence and machine learning. Most architectures of deep learning are based on neural networks, which are inspired by neuroscientific research. The earliest models of ANNs are simple linear models and associated the output value y with the input x. In other words, these models want to learn a function given a training set of samples, in which each sample is a pair of an input value and output value. For instance, the perceptron, proposed in 1957 [45], learns a linear model in a supervised way, which can be represented as $f(x, w) = wx + b$. This mathematical model mimics the operation of neurons in humans' brain. Therefore, deep learning consisting of many perceptrons can also be considered as a generalization of a linear or logistic regression and imitate the functions of brains.

From the 1980s, two technology waves for ANNs, i.e., shallow learning and deep learning, have been witnessed with the development of learning approaches. At this time, each input of models should be represented by many features, and each feature will activate a separate neuron or hidden unit. At the mid of the 1980s, the back propagation (BP) algorithm [46] was proposed to learn the parameters of artificial networks and then bring a new upsurge of statistical model-based machine learning. The artificial networks can learn statistical rules from large amounts of samples with the BP algorithm and make predictions for new instances. So the neural networks could be trained to tackle complex learning problems, e.g., handwritten zip code recognition [37]. Compared to the rule-based systems, learning-based systems, e.g., neural networks, are proved to be more competitive in real-world applications. The neural networks are also called multilayer perceptrons (MLPs); however, MLPs usually have limited hidden layers due to the difficulty of training networks even if they have large datasets.

In the early 1990s, researchers made few important advances in modeling sequences with neural networks. The main focus of neural nets was devoted to the unsupervised learning. The models of neural nets were trained to produce a low-dimensional representation of unlabeled data. Besides, the neural networks with many layers, such as 20 or more, cannot work very well in practice and suffered from the vanishing gradient problem. Consequently, a variety of shallow model-based learning approaches were proposed and popularized since 1995, such as supporting vector machine (SVM), logistic regression (LR), and random forests (RF). These shallow models can be regarded as the models with no or only one hidden layer. These approaches make a great success theoretically and practically, while neural networks keep a long silence from the mid-1990s. The shallow models are easy to train and can be used to solve the small-sized sample problem. In contrast, neural networks need tricky skills to train and cannot be analyzed theoretically. Sometimes, the neural networks are utilized as black boxes in real-world applications. Although neural networks continued to obtain good performance on some tasks, such as document recognition [38], the neural networks are too computationally costly and limited in most real-world applications.

Since 2000, as the Internet techniques were developing, those shallow models become dominated in real-world applications, such as advertisement recommendation, information searching, and spam filtering system. With the ascent of support vector machines and the failure of backpropagation, the shallow models dominate the applications in the early 2000s. Nevertheless, the term “deep learning” was firstly introduced by Aizenberg et al. in 2000 [1]. In 2006, deep learning begins to attract attentions after Hinton et al. published an article in Science [27] supported by the Canadian Institute for Advanced Research (CIFAR). In [27], the neural networks were used to learn representations from data and can be easily trained through layer-wise pre-training, which can be performed with unsupervised learning. The other CIFAR-affiliated research groups quickly showed that the same strategy could be used to train many other kinds of deep networks [6, 49]. At this time, deep network methods outperformed competing other machine learning technologies (especially shallow models) as well as hand-designed functionality, and ultimately this popularized the use of the term “deep learning.”

After the breakthrough of 2006, deep learning continues to heat up in the academic community, and several universities, e.g., Stanford University, New York University, and University of Montreal, become the centers of deep learning. In 2010, the program for deep learning was firstly supported by the Defense Advanced Research Projects Agency (DARPA) of the US government. Then, the Microsoft and Google utilized the deep learning technologies to greatly reduce the errors of speech recognition and obtained biggest breakthrough in the field of speech recognition over past 10 years [25]. In 2012, the deep learning technology was used to classify images in ImageNet challenge for the first time and dramatically improve the performance by 20%, leading to the deep learning revolution and applications in numerous domains. In other contests, e.g., robust reading challenge in International Conference on Document Analysis and Recognition (ICDAR); Microsoft image recognition, segmentation, and captioning challenge (COCO); and Face Detection Data Set and Benchmark (FDDB), the deep learning technologies have become the mainstream approaches from recent 2 years, outperforming the shallow methods based on handcrafted features. After the successful application, the focus of deep learning is on new unsupervised learning techniques at the beginning and now changed to the supervised learning algorithms for leveraging large-scale datasets.

2 Basic Types

The deep learning methods based on neural networks are consisted of multiple layers and each layer contains several units. The activation y of each unit represents a linear combination of input vector $\mathbf{x}$ and learnable parameters $\mathbf{w}$ as well as basis b, followed by an element-wise nonlinearity function $f(\cdot)$:

$$y = f(\mathbf{w}\mathbf{x} + b) \tag{1}$$

where the function $f(\cdot)$ can be a sigmoid function or restricted linear unit. The deep networks stack multiple layers with different connection structures, which are called as the *architectures*.

Various deep learning architectures, such as stacked autoencoders (SAEs), deep belief networks (DBNs), convolutional neural networks (CNNs), recurrent neural networks (RNNs), and generative adversarial networks (GANs), have been proposed and successfully used in many fields. They have been shown to achieve state-of-the-art results on many tasks [13, 36]. These architectures are fundamental ones and can be extended or combined to generate new frameworks for specific tasks. In subsequent sections, these five architectures will be introduced briefly to illustrate their differences.

2.1 Stacked Autoencoders (SAEs)

Autoencoders (AEs) are simple networks directly cascading many layers [5], which have a similar forward structure with multiple-layer perceptron (MLPs). However, the AEs learn the parameters by reconstructing the input signals precisely, while MLPs use supervised information to train models. For the sake of analysis, AEs, reconstructing the input $\mathbf{x}$ of size n on the output $\tilde{\mathbf{x}}$ with only one hidden layer $\mathbf{h}$ of size k, are shown in Fig. 1. The reconstruction is computed as follows:

$$\begin{aligned} \mathbf{h} &= f(\mathbf{W}\mathbf{x} + \mathbf{b}) \\ \tilde{\mathbf{x}} &= \tilde{f}(\tilde{\mathbf{W}}\mathbf{h} + \tilde{\mathbf{b}}) \end{aligned} \tag{2}$$

where W and $\tilde{W}$ are weight matrices. b and $\tilde{b}$ are biases. If the hidden layer $\mathbf{h}$ has the same size as the input vector and use a linear function, the model will simply learn the identity function. If the dimensionality k of $\mathbf{h}$ is smaller than n of $\mathbf{x}$, the latent representations can be learned according to the equation.

Using AEs, the original data can be projected into a lower-dimensional subspace, representing latent structures in the input. So AEs are usually used for unsupervised learning of efficient coding. Moreover, various techniques have been adopted to prevent AEs from learning the identity function and obtain many variants. For instance, the regularization or sparsity constraints can further be employed to

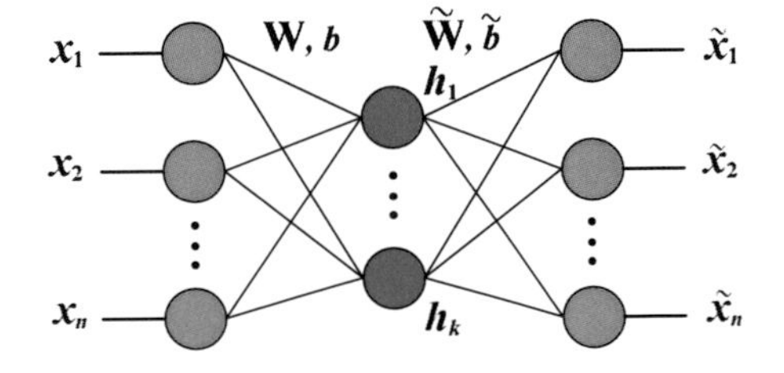

Fig. 1 The basic architecture of an AE. (©[2016] IEEE. Reprinted, with permission, from Ref. [60])

discover relevant structure more precisely and improve their ability to capture important information. In order to prevent the model from learning a trivial solution, the denoising AEs were proposed by Vincent et al. [58]. In the denoising method, the model is trained to reconstruct the input from a noise-corrupted version (typically salt-and-pepper noise).

Additionally, the SAEs are formed by stacking multiple denoising AEs layer by layer. The SAEs can be trained using gradient descent methods; however, there are fundamental problems to train deep models with many hidden layers. Once errors are backpropagated to the first few layers, they become minuscule and insignificant. Consequently, each AE is trained individually (famous as *layer-wise training*), and then the entire SAE network is fine-tuned using supervised learning technique by predicting labels given an input.

2.2 Deep Belief Networks (DBNs)

DBNs consist of many restricted Boltzmann machines (RBMs) [26], which are similar with the structure of SAEs. As shown in Fig. 2, RBMs have a similar connection way with AEs, while they utilize different optimization methods to learn parameters. Different from AEs, RBMs use bidirectional connections other than forward connections. RBMs also constitute an input layer $\mathbf{x}$ (also known as *visible layer*) and a hidden layer $\mathbf{h}$. So, essentially, RBMs are a type of Markov random field (MRF) [18]. As RBMs have bidirectional connections, they are generative models and can generate new data by the hidden layer.

Inspired by physical systems, RBMs employ an energy function as the objective function for optimization. For a state $(\mathbf{x},\mathbf{h})$, the energy function is defined as follows:

$$E(\mathbf{x},\mathbf{h}) = \mathbf{h}^T W\mathbf{x} - \mathbf{a}^T\mathbf{x} - \mathbf{b}^T\mathbf{h} \tag{3}$$

where $\mathbf{a}$ and $\mathbf{b}$ are bias terms and W is the weight matrix. Then, the probability distribution of the system over hidden and input vectors is defined by tossing the energy into an exponential function and normalizing with a partition function $\mathbf{Z}$:

$$P(\mathbf{x},\mathbf{h}) = \frac{1}{\mathbf{Z}}e^{-E(\mathbf{x},\mathbf{h})} \tag{4}$$

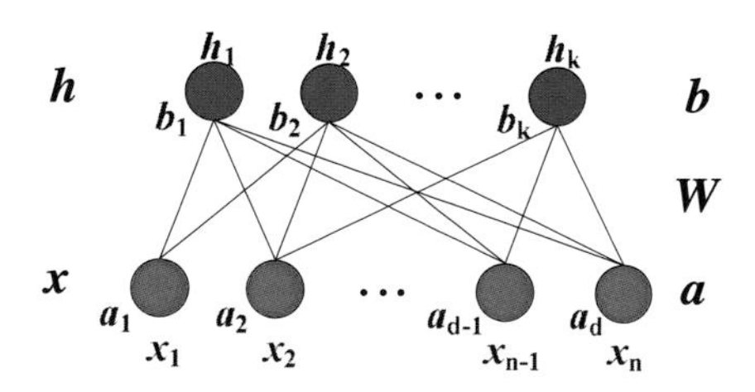

Fig. 2 The basic architecture of an RBM. (©[2016] IEEE. Reprinted, with permission, from Ref. [60])

where $\mathbf{Z}$ is defined as the sum of $e^{-E(\mathbf{x},\mathbf{h})}$ and usually difficult to compute.

Since the RBM model has bidirectional connections, the activations of hidden units are mutually independent given the activations of input units, and, conversely, the activations of input units are mutually independent given the activations of hidden units. The conditional probability of input units $\mathbf{x}$ with size of n, given the hidden units $\mathbf{h}$ with size of k, is calculated by

$$P(\mathbf{x}|\mathbf{h}) = \prod_{i=1}^{n} P(\mathbf{x}_i|\mathbf{h}) \tag{5}$$

Similarly, the conditional probability of hidden units $\mathbf{h}$ is computed by

$$P(\mathbf{h}|\mathbf{x}) = \prod_{i=1}^{k} P(\mathbf{h}_i|\mathbf{x}) \tag{6}$$

According to the conditional probability, the individual activation probabilities are given by

$$\begin{aligned} P(h_j = 1|x) &= \sigma\left(b_j + \sum_{i=1}^{n} w_{i,j} x_i\right) \\ P(x_i = 1|h) &= \sigma\left(a_i + \sum_{j=1}^{k} w_{i,j} h_j\right) \end{aligned} \tag{7}$$

where σ denotes the logistic sigmoid function.

Since DBNs stack RBMs layer by layer and have a similar structure with SAEs, the training of DBNs is performed by learning each RBM individually in an unsupervised manner and then fine-tuning on entire DBN network by adding classifiers or regressions to the last layer in a supervised manner [26]. In particular, the gradient-based contrastive divergence (CD) algorithm [7] was proposed to train deep models for DBNs, in which Gibbs sampling is used inside a gradient descent procedure to update weights.

2.3 Convolutional Neural Networks (CNNs)

SAEs and DBNs are fully connected networks as all units are connected to each other between adjacent layers. For several types of data, e.g., images, there are a large amount of parameters to learn for these fully connected networks. To avoid this problem, inspired by organization of the animal visual cortex, the convolutional neural networks (CNNs) [38] were proposed to use the weight sharing strategy for exploiting similar structures occurred in different locations in an image. Through sharing the convolutional weights locally for an entire image, this drastically reduces the amount of parameters that need to be learned and render the network equivalent with respect to translations of the input (i.e., the number of weights no longer depends on the size of input image). The basic architecture of CNNs shown in Fig. 3 contains several different layers with various functions.

The convolutional layers are the core building blocks of a CNN. At each convolutional layer, the input data is convolved with a set of K learnable kernels $W = \{W_1, W_2, \ldots, W_K\}$ adding by biases $b = \{b_1, b_2, \ldots, b_K\}$. These kernels are also known as the *receptive fields*. Then a new feature map $\mathbf{X}_k$ is generated by inputting the convolution results into an element-wise nonlinear function $\sigma(\cdot)$. Given the output vector of lth layer, the kth feature map is calculated by

$$\mathbf{X}_{\mathbf{k}}^{\mathbf{l+1}} = \sigma(W_k^l * \mathbf{X}^l + b_k^l) \tag{8}$$

The function $\sigma(\cdot)$ is also known as activation function and can be many functions, such as the sigmoid function $\sigma(x) = (1 + e^{-x})^{-1}$, hyperbolic tangent $\sigma(x) = \tanh(x)$, or rectified linear units $\sigma(x) = \max(0, x)$. Although these kernels in each convolutional layer have small receptive fields, they can be extended through the full depth of the input volume. Stacking the feature maps for all filters along the depth dimension forms the full output volume of the convolutional layer.

Usually, the pooling layer follows the convolutional layer and performs the nonlinear down-sampling. There are several nonlinear functions to implement down-sampling operations for pooling layers, among which max pooling is the most common operation. It partitions the feature map into a set of nonoverlapping rectangles and, for each such subregion, outputs the maximum values. In addition to max pooling, the pooling layer can also perform other nonlinear operations, such as average pooling or L2-norm pooling [48]. The pooling layer serves to progressively reduce the spatial size of intermediate representations, the number of parameters, and amount of computation in CNN architecture and hence to control overfitting. Besides, the pooling operation can provide a form of translation invariance.

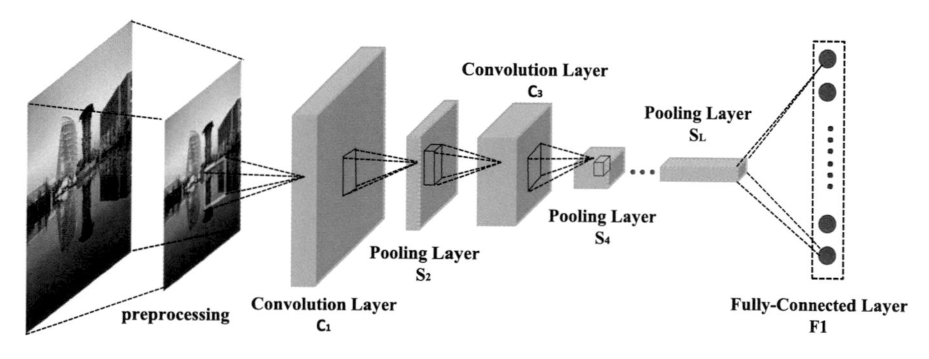

Fig. 3 The basic architecture of a CNN

Eventually, the high-level reasoning in CNNs is implemented with fully connected (FC) layers after several convolutional and pooling layers. The units in the FC layer have full connections to all activations in the previous layer, similar to SAEs and DBNs. Their activations can hence be computed with a matrix multiplication followed by a bias offset and output the classes or the probabilities. Through calculating the loss between the predicted and true data, the CNN can be trained by an end-to-end method as the number of weights is reduced by convolutional layers and pooling layers. Usually, various loss functions are adopted for different tasks, such as Softmax loss, sigmoid cross-entropy loss, and Euclidean loss.

In the early years, the first proposed architecture of CNNs was the LeNet [38], which established the foundation for subsequent architectures. In 2012, AlexNet [34] was proposed to extend LeNet with wider and more convolutional layers based on powerful GPUs. This architecture won the ImageNet Large Scale Visual Recognition Challenge (ILSVRC) and outperformed handcrafted feature-based methods. Then, the famous VGG architectures [8] were proposed to use smaller-sized filters and deeper frameworks, which can be applied to various tasks, such as image classification and face recognition. After that, the network-in-network [39] was presented to provide the great and simple insight of using 1×1 convolution to provide more combinational power to the features of convolutional layers. To go deeper layers, the GoogleNet with 22 layers [56] was proposed and made use of so-called inception blocks (Inception V2 [30], Inception V3 [56], Inception V4 [55]). Inception blocks can be regarded as a type of network-in-network and also utilize 1×1 convolution to reduce the dimensionality of the feature maps. Especially, the ResNet [23] was proposed to feed the output of two successive convolutional layers and also bypass the input to the next layers. With the shortcut connections, the architecture can use more than 1000 layers and act in parallel to serially flow the entire network [24]. In the future, much wider and deeper architectures will be proposed to achieve better performance with the developing of computation ability.

2.4 Recurrent Neural Networks (RNNs)

The aforementioned deep learning architectures, such as SAEs and CNNs, are a type of feed-forward networks, in which activations are fed forward from input to output through hidden layers. Different from them, the recurrent neural network (RNN) is a type of networks where connections between units form a directed cycle. The most basic architecture is shown in Fig. 4. Actually, more complicated connections can exist between hidden units, which are beyond the scope of this chapter. With the recurrent connections, RNNs can process sequential data, e.g., videos and speech sentences, by exhibiting dynamic temporal behaviors.

Rather than learning the posterior over $\mathbf{y}$ given single input vector $\mathbf{x}$, RNNs learn the posterior given a sequence $\mathbf{x}_1, \mathbf{x}_2, \ldots, \mathbf{x}_T$ and are more informative for sequences. The basic architecture of RNNs [16] contains the hidden layer $\mathbf{h}_t$ at time t that has the feed-forward input $\mathbf{x}_t$ and recurrent input $\mathbf{h}_{t-1}$:

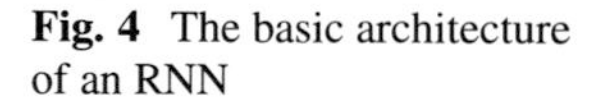
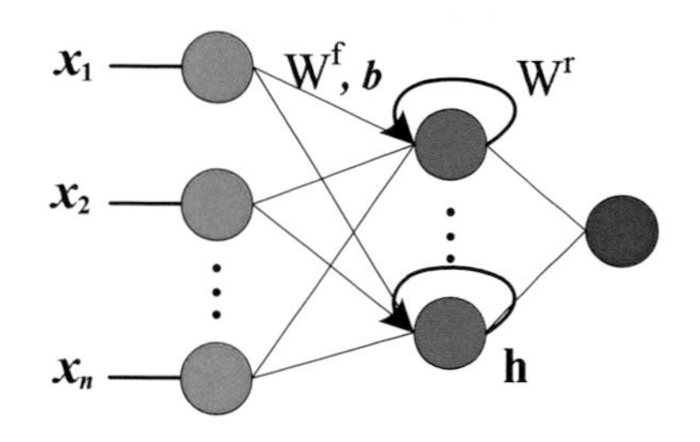

Fig. 4 The basic architecture of an RNN

$$\mathbf{h}_t = f(W^f \mathbf{x}_t + W^r \mathbf{h}_{t-1} + \mathbf{b}) \tag{9}$$

where W^f is the weight matrix for feed-forward input and W^r for recurrent input. $\mathbf{b}$ is the basis vector and $\mathbf{h}_t$ is the output of hidden units.

In the last decade, several methods for supervised training of RNNs have been presented, such as backpropagation through time (BPTT), real-time recurrent learning (RTRL), and extended Kalman filtering-based techniques (EKF). Since the gradient needs to be backpropagated from the output through time, RNNs are inherently deep in time and consequently suffer from the problems of fading or exploding gradients during learning. To solve the vanishing gradient problem, the memory units can be added into the traditional RNNs. The most common one is the long short-term memory (LSTM) [28]. The LSTM is normally augmented by recurrent gates called forget gates and can prevent back-propagated errors from vanishing or exploding. Each gate is governed by a weight matrix from the input and a weight matrix from the previous hidden layer. At the heart of the unit lies a memory cell that combines the activation of the other gates and relays it to the output of the unit and the next state of its memory. Instead errors can flow back-forward through unlimited numbers of virtual layers in LSTM RNNs unfolded in space.

Aforementioned recurrent networks have a causal structure that the output depends on the past and present input. However, in many applications, e.g., speech recognition, the output may depend on the whole input sequence. To solve this problem, the bidirectional RNNs [50] integrated two RNNs in one network: an RNN that moves forward through time beginning from the start of a sequence and the other RNN that moves backward through time beginning from the end of the sequence.

2.5 Generative Adversarial Nets (GANs)

The basic idea of generative adversarial networks (GANs) [20] is that it contains two sub-models: a generative model and a discriminative model. The basic architecture is shown in Fig. 5. The discriminative model in terms of discriminator has a task of determining whether a given sample looks natural (a sample from the dataset) or looks that it has been artificially created. The task of generative model having

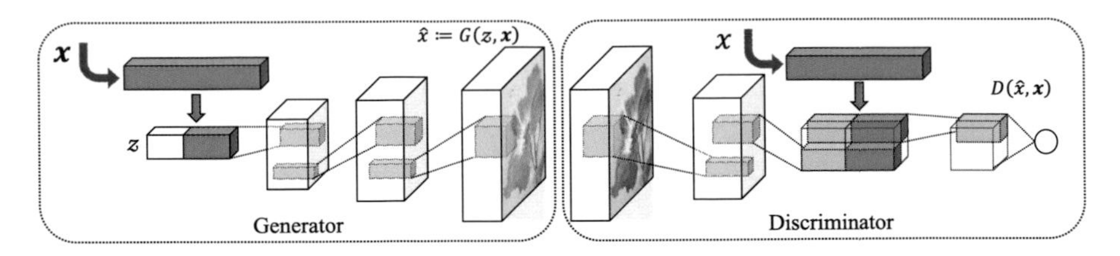

Fig. 5 The basic architecture of a GAN

a name of generator is to create natural-looking samples which are similar to the original data distribution. This can be thought of as a zero-sum or minimax two-player game.

The generator's distribution p_g over data $\mathbf{x}$ needs to be learned given a prior $p_z(\mathbf{z})$ on noise variables $\mathbf{z}$, which are mapped to data space as $G(\mathbf{z}; \theta_g)$. The generator G is a differentiable function represented by a deep model, such as a multilayer perceptron or convolutional network. The discriminator D outputs the probability that $\mathbf{x}$ comes from the original data rather than p_g. The discriminator D can also be represented as a differentiable function that has similar deep structure with the generator. It can be trained by maximizing the probability of assigning the correct labels to both training examples and generated samples. Simultaneously, the generator is trained by minimizing the objective function $\log(1 - D(G(\mathbf{z})))$. The optimal function for generator and discriminator can be defined as follows:

$$\min_G \max_D \mathbb{E}_{x \sim p_x}[\log D(x)] + \mathbb{E}_{z \sim p_z(\mathbf{z})}[\log(1 - D(G(\mathbf{z})))] \tag{10}$$

The theoretical analysis of GANs demonstrates that the training criterion allows one to recover the data generating distribution [20]. In practice, the optimization is implemented with an iterative, numerical approach. According to Eq. 10, optimizing D in the inner loop of training is computationally prohibitive and would induce the problem of overfitting on finite training samples. Instead, it is alternative to change between k steps of optimizing D and one step of optimizing G. However, the training method faces several challenges, such as non-convergence and collapse problem.

To solve the problem of training instability, the Wasserstein GANs [22] were recently proposed toward stable training of GANs. This model presented several ingenious skills to improve the training of GANs. The use of weight clipping was changed in Wasserstein GANs to enforce the Lipschitz constraint, and instead the norm of the gradient was penalized. This method converges faster and generates higher-quality samples than original GAN with weight clipping.

3 Practical Applications

Deep learning techniques have been widely applied in many areas, e.g., computer vision and natural language processing, and are becoming the state-of-the-art methods in these domains. According to the types of data, the application domains of deep learning will be categorized by audio data, image data, and text data. Toward the specific tasks, various deep learning approaches have been presented to greatly expand the basic types of deep learning. More specifically, these applications of deep learning refer to many hot topics, such as speech recognition, face recognition, object detection, scene classification, and machine translation.

3.1 Audio Data

Audio data comes in a variety of forms, in which the speech is the most common one. In the past decades, the speech recognition system mostly adopts the Gaussian mixture model (GMM) to describe the statistical probability models of each modeling unit. This model has long been occupied monopoly position in speech recognition applications due to its simple estimation for mass data as well as the matured techniques of discrimination training. However, the GMM is essentially a shallow network and cannot fully describe the characteristics of the distribution of state space. In addition, the feature dimensionality of GMM modeling is generally dozens and makes it difficult to describe the correlation between features. Eventually, the GMM modeling is essentially a likelihood probability model. Although the partition training can mimic the distinction between some classes, the ability is still limited [25].

In 2011, Microsoft first presented a speech recognition system integrating DBNs method and completely changed the original framework of speech recognition technology [11]. After that, other corporations, such as Google, IBM, Baidu, and iFlytek, began to use the deep learning technologies to model the speech [25, 31]. With the use of deep neural networks, the correlation between features of speech can be fully described, and continuous multi-frame speech features can be combined to constitute a high-dimensional feature. The final neural networks can be trained by this high-dimensional feature. Since the multilayer neural networks are used to simulate the human brain, the feature extraction can be performed layer by layer, and ultimately the formation of ideal features can be obtained for speech classification. The multilayer structure is similar to the processing of human for speech recognition. Recently, the LSTM-based RNNs [21] have been employed to avoid the vanishing gradient problem and can learn very deep architectures specially for speech, which is now available through Google Voice to all mobile phone users.

On the other side, the deep learning technologies can seamlessly be integrated into the traditional speech recognition systems without causing any additional cost in case of greatly increased recognition rate. The integration method can be implemented by replacing the likelihood of hidden Markov models (HMM) with the output likelihood of deep learning, while the traditional speech model, statistical language model, and dynamic decoder are still used in the process of decoding. In practice, compared to speech recognition system using traditional HMM models, deep learning-based speech recognition system improved the recognition rate by at least 30% [11]. Consequently, deep learning makes the speech recognition system widely applicable and fully effects the system in the future [12].

Besides, Zen et al. [61] presented a speech synthesis model based on multilayer perceptron. In this model, the first input voice is changed into an input feature sequence, and then each frame of input sequences is mapped to the output feature, respectively, through multilayer perceptron generating speech parameters. Moreover, the deep learning was applied to music retrieval, and the 1-D CNNs were presented to implement the end-to-end learning for music audios [14]. Similar to music retrieval, CNNs were also applied to recommend music based on deep content [43], in which a dense neural net layer is trained to predict the latent factors.

3.2 Image Data

In 2012, the CNNs were firstly used to ImageNet challenge for image classification and obtained best performance by Krizhevsky et al. [34]. In subsequent years, many CNN-based deep methods have been presented in the challenge and achieved state-of-the-art results. Moreover, the strategy of multiple sliding windows [51] is combined with CNNs to localize objects in the challenge, and deeper architectures with more traditional strategies have been applied to this task [44]. From 2014, all teams in the challenge have adopted deep learning methods to compete with each other [47]. Compared with the traditional approaches, the deep learning techniques occupy a great advantage in the tasks of image classification and object detection.

Another successful application for image data is the task of face verification. In 2014, the DeepID project [54] from the Chinese University of Hong Kong (CUHK) and DeepFace project [57] from Facebook employed CNNs to recognize faces in the wild, respectively. They have achieved recognition precisions by 97.45% and 97.35% on the labeled faces in the wild (LFW) database. Then, the DeepID2 [53] (updated version of DeepID) from CUHK achieved 99.15% recognition precision, better than the humans' performance, and outperformed all existing approaches. Recently, several startups have reported better performances with ResNet-based deep modes, such Face++ and AuthenMetric [35]. These deep models employ similar convolutional layers and fully connected layers, while they have different depths and widths.

For large-scale video classification, the CNNs have been used by Karpathy et al. to classify 1 million YouTube videos [33]. In this deep model, the temporal information has been leveraged to construct multiple networks with fusing single frames and multiple frames. The classification accuracy has been promoted by at least 8%. Then, the RNNs with LSTM cells [42] were employed to model videos as an ordered sequence, in which the performance was improved about 10%. Currently, the deep learning becomes the mainstream methods for classifying YouTube-8M dataset on Kaggle challenge and continues to be promoted for real-world applications.

The CNN architecture and its variants are also applied to the field of human action recognition. A deep learning method [2] that is sequential was presented to classify human behaviors without any prior knowledge. The first step is to extend the traditional convolutional networks to three-dimensionality and then automatically learn the spatial and temporal characteristics of action videos. Then it used the RNN method to train the classification of each sequence. The results of deep model on KTH dataset are better than other known deep models and obtain the accuracy 94.39% and 92.17% on KTH1 and KTH2, respectively. Another 3D convolutional network [32] was proposed to extract spatial and temporal features for representing motion information of multiple frames. This approach generated multiple feature channels and fused them for final representations. The performance of 3D model outperformed other traditional methods on TRECVID dataset.

In recent two years, the deep learning techniques have been applied to almost all kinds of image data, such as medical images [40], remote sensing images [62], and natural images. Other applications for image data refer to image super-resolution reconstruction [15], image fusion [41], image hashing [60], scene parsing [17], texture recognition[3], and doorplate number recognition [19]. In the future, more and more topics concerned with image data will be affected by deep learning and outperform the humans' performance.

3.3 *Text Data*

In the fields of natural language processing (NLP), text mining, and information retrieval (IR), their subtasks are concerned with the text data, which are usually discrete and sequential. Therefore, 1D convolutional and recurrent structures are more adaptive to text data, which are different from audio and image data. The deep learning techniques can be used to derive high-quality information and obtain appreciated representations and leverage the structures of text data.

The first deep learning framework was applied to classical NLP problems by Collobert et al. in 2008 [10]. This framework used 1D convolutional networks for all subtasks and achieved state-of-the-art performances. Then, Cho et al. [9] presented an RNN-based framework with fixed-size vector representations for machine translation. In this deep model, two RNNs are included: one RNN is used to encode a set of source language symbol sequences into a set of fixed-size vectors, and the other RNN decodes the vector into a set of symbolic sequences of the target language. To overcome the disadvantages of fixed-size vector representations, Bahdanau et al. [4] proposed a deep model based on bidirectional RNNs. In the translation of each word, this model predicts target words according to the most relevant information of words in the source text and the location of other translated words. Besides, this model contains a bidirectional RNN as an encoder and a decoder for word translation. In the prediction of target word position, a multilayer perceptron model is used for position alignment.

For the task of information retrieval, the deep learning can be used to represent query texts and documents. Furthermore, the matching and ranking procedure can employ deep learning methods to calculate the complicated similarities between question and fact in knowledge base. Huang et al. [29] applied the deep structure learning method to enhance the relevance for matching, while CNNs were used to learn to rank by Severyn et al. [52]. In real-world applications, the deep learning techniques have been successfully used in commercial search engines, e.g., Google and Baidu, for promoting the search performance. Moreover, using retrieval-based deep learning techniques, the application of question answering has been promoted to return answers from a large repository of question answer pairs [59]. In the future, deep learning will be particularly effective for more IR problems.

4 Existing Challenges

4.1 Theory Challenges

Compared to other learning methods, such as SVM, the deep learning techniques are difficult to interpret theoretically. As the deep models are usually consisted of several layers and amounts of units, it is impossible to interpret local structure individually. Sometimes, the deep learning methods are regarded as black boxes. Since the deep models have thousands of parameters, the convergence of training strategies cannot be analyzed easily. Especially, an important deficiency is that the parameter initialization of deep learning needs more precise guidance in theory.

Compared to the shallow models, the deep models have more powerful ability of representations with the nonlinear function. Usually, for an arbitrary nonlinear function, both shallow networks and deep networks can be found for good representations according to the universal approximation theory of neural networks. However, for several functions, the deep networks only require far fewer parameters. So, we need to understand the sample complexity that is how many samples we need to learn a good deep model. It is very difficult to study this theoretically because of the non-convex function.

On the other side, it is still difficult to design a deep model for specific tasks. It seems to be that the common processing unit, like the convolution, exits in deep models for audio, image, and text data. It is interesting that whether there can establish a unified framework for any kind of data. Besides, in specific tasks, it is still a challenge to incorporate domain expertise in these deep models.

In addition, it is still a problem that uses the deep model to represent the structured semantic information. From an evolutionary point of view, language ability is far behind the visual and auditory development. So from this perspective, it is difficult to process abstract problems of language, and successfully solving this problem can make the realization of artificial general intelligence.

4.2 *Engineering Challenges*

In the real-world applications, large-scale parallel computing platform becomes the first engineering problem to perform massive data training for deep learning technology. The traditional big data processing platforms, such as Hadoop and Spark, are not suitable for deep learning as they have high latency. The existing deep learning technologies usually use the stochastic gradient method (SGD) method to train. This method itself cannot be parallel between multiple computers. Even using GPUs, the training time of traditional deep models is still very long. The traditional CNN model still needs hundreds of hours. With the Internet service development, the training data is more and more important, and the slow training speed cannot meet the needs of Internet service application. Parallel computing platforms will overcome the technical problems of traditional SGD method. With the development of hardware and platforms, it can be expected in the future that massive training data will continue to improve the ability of deep learning.

For various tasks, it is challenging to train deep models for high-performance requirement. The difficulties usually emerge in two aspects. Firstly, the deep architectures have hundreds of thousands of parameters to train. To obtain these optimal values, several special skills for training deep models need to be found, while there do not exist universal training skills for all tasks. For every individual task, the processing of data and parameters are unique and different from others. Secondly, the deep models need large-scale samples to learn, while many tasks cannot provide enough samples. Especially, it is costly for some applications, e.g., medical image analysis, to obtain high-quality supervision information. So the insufficient samples with high-quality supervisions limit the practical application of deep learning.

5 Conclusions

Deep learning has brought a new wave of machine learning and artificial intelligence and attracted wide attention from academia to industry. This has also led to the era of big data with deep learning. In theories, various methods for deep learning have been proposed for different tasks. In applications, intelligent recognition and understanding of speech, image, and text have made amazing progresses to promote artificial intelligence and human-computer interaction. At the same time, complex systems with the task of learning have also been significantly improved. In a word, it is the era of deep learning.

References

1. Aizenberg, I.N., Aizenberg, N.N., Vandewalle, J.P.: Multi-valued and universal binary neurons: Theory, learning and applications. Springer US (2000)
2. Baccouche, M., Mamalet, F., Wolf, C., Garcia, C., Baskurt, A.: Sequential deep learning for human action recognition. In: International Conference on Human Behavior Understanding, pp. 29–39 (2011)
3. Badri, H., Yahia, H., Daoudi, K.: Fast and accurate texture recognition with multilayer convolution and multifractal analysis. In: European Conference on Computer Vision (ECCV), pp. 505–519 (2014)
4. Bahdanau, D., Cho, K., Bengio, Y.: Neural machine translation by jointly learning to align and translate. arXiv (2014)
5. Bengio, Y., Lamblin, P., Popovici, D., Larochelle, H.: Greedy layer-wise training of deep networks. In: Advances in Neural Information Processing Systems (NIPS), pp. 153–160 (2007)
6. Bengio, Y., Senecal, J.S.: Adaptive importance sampling to accelerate training of a neural probabilistic language model. IEEE Transactions on Neural Networks **19**(4), 713–22 (2008)
7. Carreira-Perpinan, M.A., Hinton, G.E.: On contrastive divergence learning. Artificial Intelligence & Statistics (2005)
8. Chatfield, K., Simonyan, K., Vedaldi, A., Zisserman, A.: Return of the devil in the details: Delving deep into convolutional nets. In: British Machine Vision Conference (BMVC), pp. 1–12 (2014)
9. Cho, K., Merrienboer, B.V., Gulcehre, C., Bahdanau, D., Bougares, F., Schwenk, H., Bengio, Y.: Learning phrase representations using RNN encoder-decoder for statistical machine translation. arXiv (2014)
10. Collobert, R., Weston, J., Bottou, L., Karlen, M., Kavukcuoglu, K., Kuksa, P.: Natural language processing (almost) from scratch. Journal of Machine Learning Research **12**(1), 2493–2537 (2011)
11. Dahl, G.E., Yu, D., Deng, L., Acero, A.: Context-dependent pre-trained deep neural networks for large-vocabulary speech recognition. IEEE Transactions on Audio Speech & Language Processing **20**(1), 30–42 (2012)
12. Deng, L., Hinton, G., Kingsbury, B.: New types of deep neural network learning for speech recognition and related applications: An overview. In: IEEE Conference on Acoustics, Speech and Signal Processing (ICASSP), pp. 8599–8603 (2013)
13. Deng, L., Yu, D.: Deep learning: Methods and applications. Foundations & Trends in Signal Processing **7**(3), 197–387 (2013)
14. Dieleman, S., Schrauwen, B.: End-to-end learning for music audio. In: IEEE Conference on Acoustics, Speech and Signal Processing (ICASSP), pp. 6964–6968 (2014)
15. Dong, C., Chen, C.L., He, K., Tang, X.: Learning a deep convolutional network for image super-resolution. In: European Conference on Computer Vision (ECCV), vol. 8692, pp. 184–199 (2014)
16. Elman, J.L.: Finding structure in time. Cognitive Science **14**(2), 179–211 (1990)
17. Farabet, C., Couprie, C., Najman, L., Lecun, Y.: Learning hierarchical features for scene labeling. IEEE Transactions on Pattern Analysis & Machine Intelligence **35**(8), 1915–1929 (2013)
18. Fischer, A., Igel, C.: Training restricted Boltzmann machines: An introduction. Pattern Recognition **47**(1), 25–39 (2014)
19. Goodfellow, I.J., Bulatov, Y., Ibarz, J., Arnoud, S., Shet, V.: Multi-digit number recognition from street view imagery using deep convolutional neural networks. arXiv (2014)
20. Goodfellow, I.J., Pougetabadie, J., Mirza, M., Xu, B., Wardefarley, D., Ozair, S., Courville, A., Bengio, Y., Ghahramani, Z., Welling, M.: Generative adversarial nets. In: Advances in Neural Information Processing Systems (NIPS), vol. 3, pp. 2672–2680 (2014)
21. Graves, A., Jaitly, N.: Towards end-to-end speech recognition with recurrent neural networks. In: International Conference on Machine Learning (ICML), pp. 1764–1772 (2014)

22. Gulrajani, I., Ahmed, F., Arjovsky, M., Dumoulin, V., Courville, A.: Improved training of wasserstein GANs. arXiv (2017)
23. He, K., Zhang, X., Ren, S., Sun, J.: Deep residual learning for image recognition. In: IEEE Conference on Computer Vision and Pattern Recognition (CVPR), pp. 770–778 (2016)
24. He, K., Zhang, X., Ren, S., Sun, J.: Identity mappings in deep residual networks. In: European Conference on Computer Vision (ECCV) (2016)
25. Hinton, G., Deng, L., Yu, D., Dahl, G.E.: Deep neural networks for acoustic modeling in speech recognition: The shared views of four research groups. IEEE Signal Processing Magazine **29**(6), 82–97 (2012)
26. Hinton, G.E.: A practical guide to training restricted Boltzmann machines. Momentum **9**(1), 599–619 (2012)
27. Hinton, G.E., Salakhutdinov, R.R.: Reducing the dimensionality of data with neural networks. Science **313**(5786), 504 (2006)
28. Hochreiter, S., Schmidhuber, J.: Long short-term memory. Neural Computation **9**(8), 1735 (1997)
29. Huang, P.S., He, X., Gao, J., Deng, L., Acero, A., Heck, L.: Learning deep structured semantic models for web search using clickthrough data. In: ACM Conference on Conference on Information & Knowledge Management, pp. 2333–2338 (2013)
30. Ioffe, S., Szegedy, C.: Batch normalization: Accelerating deep network training by reducing internal covariate shift. arXiv (2015)
31. Jaitly, N., Nguyen, P., Senior, A., Vanhoucke, V.: Application of pretrained deep neural networks to large vocabulary conversational speech recognition. In: Proceedings of Interspeech (2012)
32. Ji, S., Yang, M., Yu, K.: 3d convolutional neural networks for human action recognition. IEEE Transactions on Pattern Analysis & Machine Intelligence **35**(1), 221–231 (2013)
33. Karpathy, A., Toderici, G., Shetty, S., Leung, T.: Large-scale video classification with convolutional neural networks. In: IEEE Conference on Computer Vision and Pattern Recognition (CVPR), pp. 1725–1732 (2014)
34. Krizhevsky, A., Sutskever, I., Hinton, G.E.: Imagenet classification with deep convolutional neural networks. In: Advances in Neural Information Processing Systems (NIPS), pp. 1097–1105 (2012)
35. Learned-Miller, E., Huang, G.B., Roychowdhury, A., Li, H., Hua, G.: Labeled faces in the wild: A survey. Springer International Publishing (2016)
36. Lecun, Y., Bengio, Y., Hinton, G.: Deep learning. Nature **521**(7553), 436–444 (2015)
37. Lecun, Y., Boser, B., Denker, J.S., Henderson, D., Howard, R.E., Hubbard, W., Jackel, L.D.: Backpropagation applied to handwritten zip code recognition. Neural Computation **1**(4), 541–551 (1989)
38. Lecun, Y., Bottou, L., Bengio, Y., Haffner, P.: Gradient-based learning applied to document recognition. Proceedings of the IEEE **86**(11), 2278–2324 (1998)
39. Lin, M., Chen, Q., Yan, S.: Network in network. In: International Conference on Learning Representations (ICLR) (2014)
40. Litjens, G., Kooi, T., Bejnordi, B.E., Setio, A.A.A., Ciompi, F., Ghafoorian, M., van der Laak, J.A.W.M., Van Ginneken, B., Snchez, C.I.: A survey on deep learning in medical image analysis. arXiv (2017)
41. Masi, G., Cozzolino, D., Verdoliva, L., Scarpa, G.: Pansharpening by convolutional neural networks. Remote Sensing **8**(7), 594 (2016)
42. Ng, Y.H., Hausknecht, M., Vijayanarasimhan, S., Vinyals, O.: Beyond short snippets: Deep networks for video classification. In: IEEE Conference on Computer Vision and Pattern Recognition (CVPR), vol. 16, pp. 4694–4702 (2015)
43. Oord, A.V.D., Dieleman, S., Schrauwen, B.: Deep content-based music recommendation. In: Advances in Neural Information Processing Systems Conference (NIPS), pp. 2643–2651 (2013)

44. Ren, S., He, K., Girshick, R., Sun, J.: Faster R-CNN: Towards real-time object detection with region proposal networks. IEEE Transactions on Pattern Analysis & Machine Intelligence pp. 1–14 (2016)
45. Rosenblatt, F.: The perceptron – a perceiving and recognizing automaton. In: Math. Stat (1957)
46. Rumelhart, D.E., Hinton, G.E., Williams, R.J.: Learning representations by back-propagating errors. Nature **323**(6088), 533–536 (1986)
47. Russakovsky, O., Deng, J., Su, H., Krause, J., Satheesh, S., Ma, S., Huang, Z., Karpathy, A., Khosla, A., Bernstein, M.: Imagenet large scale visual recognition challenge. International Journal of Computer Vision **115**(3), 211–252 (2015)
48. Scherer, D., Ller, A., Behnke, S.: Evaluation of pooling operations in convolutional architectures for object recognition. In: International Conference on Artificial Neural Networks, pp. 92–101 (2010)
49. Schlkopf, B., Platt, J., Hofmann, T.: Efficient learning of sparse representations with an energy-based model. In: Advances in Neural Information Processing Systems (NIPS), pp. 1137–1144 (2006)
50. Schuster, M., Paliwal, K.K.: Bidirectional recurrent neural networks. IEEE Transactions on Signal Processing **45**(11), 2673–2681 (1997)
51. Sermanet, P., Eigen, D., Zhang, X., Mathieu, M., Fergus, R., Lecun, Y.: Overfeat: Integrated recognition, localization and detection using convolutional networks. arXiv (2013)
52. Severyn, A., Moschitti, A.: Learning to rank short text pairs with convolutional deep neural networks. In: The International ACM SIGIR Conference, pp. 373–382 (2015)
53. Sun, Y., Wang, X., Tang, X.: Deep learning face representation by joint identification-verification. In: Advances in Neural Information Processing Systems (NIPS), vol. 27, pp. 1988–1996 (2014)
54. Sun, Y., Wang, X., Tang, X.: Deep learning face representation from predicting 10,000 classes. In: IEEE Conference on Computer Vision and Pattern Recognition (CVPR), pp. 1891–1898 (2014)
55. Szegedy, C., Ioffe, S., Vanhoucke, V., Alemi, A.: Inception-v4, inception-resnet and the impact of residual connections on learning. arXiv (2016)
56. Szegedy, C., Liu, W., Jia, Y., Sermanet, P.: Going deeper with convolutions. In: IEEE Conference on Computer Vision and Pattern Recognition (CVPR), pp. 1–9 (2015)
57. Taigman, Y., Yang, M., Ranzato, M., Wolf, L.: Deepface: Closing the gap to human-level performance in face verification. In: IEEE Conference on Computer Vision and Pattern Recognition (CVPR), pp. 1701–1708 (2014)
58. Vincent, P., Larochelle, H., Lajoie, I., Bengio, Y., Manzagol, P.A.: Stacked denoising autoencoders: Learning useful representations in a deep network with a local denoising criterion. Journal of Machine Learning Research **11**(12), 3371–3408 (2010)
59. Wang, M., Lu, Z., Li, H., Liu, Q.: Syntax-based deep matching of short texts. In: International Conference on Artificial Intelligence, pp. 1354–1361 (2015)
60. Xia, Z., Feng, X., Peng, J., Hadid, A.: Unsupervised deep hashing for large-scale visual search. In: International Conference on Image Processing Theory, Tools and Applications (IPTA) (2016)
61. Ze, H., Senior, A., Schuster, M.: Statistical parametric speech synthesis using deep neural networks. In: IEEE Conference on Acoustics, Speech and Signal Processing (ICASSP), pp. 7962–7966 (2013)
62. Zhang, L., Zhang, L., Du, B.: Deep learning for remote sensing data: A technical tutorial on the state of the art. IEEE Geoscience & Remote Sensing Magazine **4**(2), 22–40 (2016)

Deep Learning in Object Detection

Yanwei Pang and Jiale Cao

Abstract Object detection is an important research area in image processing and computer vision. The performance of object detection has significantly improved through applying deep learning technology. Among these methods, convolutional neural network (CNN)-based methods are most frequently used. CNN methods mainly include two classes: two-stage methods and one-stage methods. This chapter firstly introduces some typical CNN-based architectures in details. After that, pedestrian detection, as a classical subset of object detection, is further introduced. According to whether CNN is used or not, pedestrian detection can be divided into two types: handcrafted feature-based methods and CNN-based methods. Among these methods, NNNF (non-neighboring and neighboring features) inspired by pedestrian attributes (i.e., appearance constancy and shape symmetry) and MCF based on handcrafted channels and each layer of CNN are specifically illustrated. Finally, some challenges of object detection (i.e., scale variation, occlusion, and deformation) will be discussed.

1 Introduction

Object detection can be applied into many computer vision areas, such as video surveillance, robotics, and human interaction. However, due to the factors of complex background, illumination variation, scale variation, occlusion, and object deformation, object detection is very challenging and difficult. In the past few decades, researchers have done a lot of work to push the progress of object detection. Figure 1 shows mean average precisions (mAP) on PASCAL VOC2007 of some methods (i.e., [18, 23–25, 51, 52, 63]).

Depending on whether deep learning is used or not, object detection methods can be divided into two main classes: the handcrafted feature-based methods

Y. Pang (✉) · J. Cao
School of Electrical and Information Engineering, Tianjin University, Tianjin, China
e-mail: pyw@pyw.edu.cn; connor@tju.edu.cn

X. Jiang et al. (eds.), *Deep Learning in Object Detection and Recognition*,
https://doi.org/10.1007/978-981-10-5152-4_2

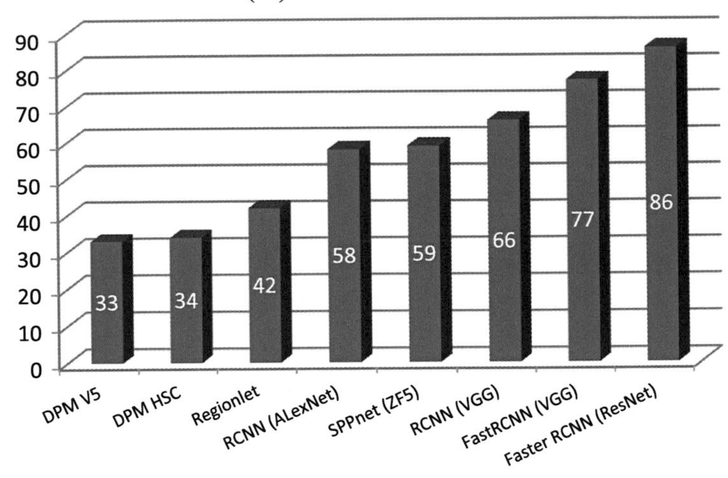

Fig. 1 The progress of object detection on PASCAL VOC2007

[6, 14, 63, 73] and the deep learning-based methods [23, 25, 36, 38, 52]. In the first decade of twenty-first century, the traditional handcrafted feature-based methods are main stream. During this period, many famous and successful image feature descriptors (e.g., SIFT [39], HOG [10], Haar [61], LBP [44], and DPM [19]) are proposed. Based on these feature descriptors and the classical classifiers (e.g., SVM and AdaBoost), these methods achieve great success at that time.

However, when it comes to 2010, the performance of objection detection tends to be stable. Though many methods are still proposed, the performance improvement is relatively limited. Meanwhile, deep learning begins to show superior performance on some computer vision areas (e.g., image classification [26, 33, 56, 57]). In 2012, with the big image data (i.e., ImageNet [13]), deep CNN network (called AlexNet [33]) achieves the best detection performance on ImageNet ILSVRC2012, which outperforms the second best method by 10.9% on ILSVRC-2012 competition.

With the great success of deep learning on image classification [13, 26, 56, 57], researchers start to explore how to improve object detection performance with deep learning. In the recent few years, object detection based on deep learning has also achieved a great progress [23, 25, 52]. The mAP of object detection on PASCAL VOC2007 [17] dramatically increases from 58% (based on RCNN with AlexNet [23]) to 86% (based on Faster RCNN with ResNet [26]). Currently, the state-of-the-art methods for deep object detection are based on deep convolutional neural networks (CNN) [23, 27, 36, 52].

In Sect. 2, some typical CNN architectures of object detection will be introduced. Pedestrian detection, as a special case of object detection, will be specifically discussed in Sect. 3. Finally, some representative challenges (i.e., occlusion, scale variation, and deformation) of object detection will be illustrated in Sect. 4.

2 The CNN Architectures of Object Detection

According to the pipeline of deep object detection, the methods can be divided into two main classes in Fig. 2: two-stage methods [11, 23–25, 52] and one-stage methods [36, 38, 49, 54]. Two-stage methods firstly generate some candidate object proposals and then classify these proposals into the specific categories. One-stage methods simultaneously extract and classify all the object proposals. Generally speaking, two-stage methods have a relatively slower detection speed and higher detection accuracy, while one-stage methods have a much faster detection speed and comparable detection accuracy. In the following part of this section, two-stage methods and one-stage methods are introduced, respectively.

2.1 Two-Stage Methods for Deep Object Detection

Two-stage methods treat object detection as a multistage process. Given an input image, some proposals of possible objects are firstly extracted. After that, these proposals are further classified into the specific object categories by the trained classifier. The benefits of these methods can be summarized as follows: (1) It reduces a large number of proposals which are put into the following classifier. Thus, it can accelerate detection speed. (2) The step of proposal generation can be seen as a bootstrap technique. Based on the proposals of possible objects, the classifier can focus on the classification task with little influence of background (or easy negatives) in the training stage. Thus, it can improve detection accuracy. Among these two-stage methods, the series of RCNN, including RCNN [23], SPPnet [25], Fast RCNN [24], and Faster RCNN [52], are very representative.

With the great success of deep convolutional neural networks (CNN) on image classification [33, 56], Girshick et al. [23] initially attempted to apply deep CNN to object detection and proposed RCNN. Compared to the traditional highly tuned

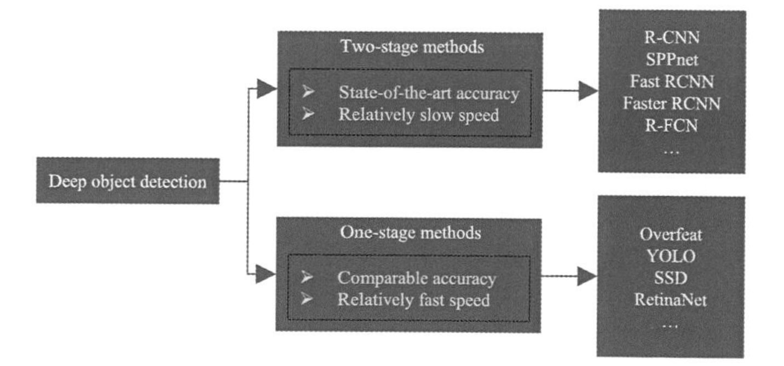

Fig. 2 Object detection in deep learning can be mainly divided into two different classes: two-stage methods and one-stage methods

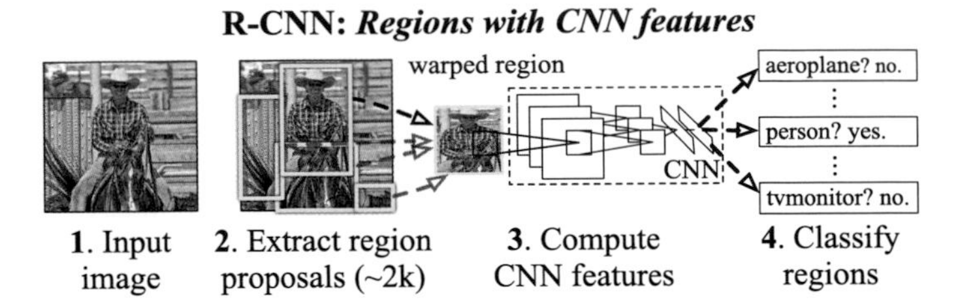

Fig. 3 The architecture of RCNN. It consists of three steps: proposal generation, CNN feature computation, and proposal classification. (©[2014] IEEE. Reprinted, with permission, from Ref. [23])

DPM [19], RCNN improves mean average precision (mAP) by 21% on PASCAL VOC2010. Figure 3 shows the architecture of RCNN. It can be mainly divided into three steps: (1) It firstly extracts the candidate object proposals, where the object proposals are category-independent. They can be extracted by the objectness methods, such as selective search [60], EdgeBox [75], and BING [9]. (2) For each object proposal of arbitrary scale, the image data is then warped into a fixed size (e.g., 227×227) and put into the deep CNN network (e.g., AlexNet [33]) to compute a 4096-d feature vector. (3) Finally, based on the feature vector extracted by CNN network, the SVM classifiers predict the specific category of each proposal.

In the training stage, the object proposals should be firstly generated by selective search for training CNN network and SVM classifiers. The CNN network (e.g., AlexNet [33]) is firstly pre-trained on ImageNet [13] and then fine-tuned on specific object detection dataset (e.g., PASCAL VOC [17]). Because the number of object category on ImageNet [13] and PASCAL VOC [17] is different, the outputs of final fully connected layer in CNN network should be changed from 1000 to 21 when fine-tuning on PASCAL VOC. The number of 21 represents 20 object classes of PASCAL VOC and the background. When fine-tuning the CNN network, the proposal is labelled as the positive for the matched class if it has the maximum IoU overlap with a ground-truth bounding box and the overlap is at least 0.5. Otherwise, the proposal is labelled as the background class. Based on the CNN features extracted from the trained CNN network, the linear SVM classifiers for different classes are further trained, respectively. When training the SVM classifier per class, only the ground-truth bounding box is labelled as the positive. Otherwise, the proposal is labelled as the negative if it has the IoU overlap below 0.3 with all the ground-truth bounding boxes. Because the extracted CNN features are too large to load in memory, the bootstrap technique is used to mine the hard negatives in training SVM classifiers.

To improve location accuracy of the proposal, the linear regression model is further trained to predict a more accurate bounding box based on the *pool*5 features of trained CNN network (i.e., AlexNet [33]). Assuming that the original bounding box of the proposal (i.e., P) is represented by (P_x, P_y, P_w, P_h), where P_x and P_y are the coordinates of the center of the proposal P, P_w and P_h are the width and the height of the proposal P. The bounding box of corresponding ground-truth (i.e., G) is represented by (G_x, G_y, G_w, G_h). Then, the regression target for a given (P, G) can be written as:

$$t_x = (G_x - P_x)/P_w, \quad (1)$$

$$t_y = (G_y - P_y)/P_h, \quad (2)$$

$$t_w = \log(G_w/P_w), \quad (3)$$

$$t_h = \log(G_h/P_h). \quad (4)$$

To predict the regression target (i.e., $(\hat{t}_x, \hat{t}_y, \hat{t}_w, \hat{t}_h)$) for a new proposal (i.e., P), the *pool*5 features of proposal represented as $\phi_5(P)$ are used. Thus, $\hat{t}_*(P) = \mathbf{w}_*{}^T \phi_5(P)$, where $\mathbf{w}_*$ is a learnable parameter and $*$ means one of x, y, w, h. Given the training sample pairs $\{(P_i, G_i)\}$, where $i = 1, 2, \ldots, N$. $\mathbf{w}_*$ can be optimized by the regularized least squares objective as follows:

$$\mathbf{w}_* = \arg\min_{\hat{\mathbf{w}}_*} \sum_i^N (\hat{t}_* - \hat{\mathbf{w}}_*^T \phi_5(P))^2 + \lambda||\hat{\mathbf{w}}_*||^2, \quad (5)$$

where λ is a regularization factor, which is usually set as 1000. When training the regression model per class, the proposal that has an IoU overlap over 0.6 with a ground-truth bounding box is used. Otherwise, the proposal is ignored.

Based on the learned $\mathbf{w}_*$ (i.e., $\mathbf{w}_x$,$\mathbf{w}_y$,$\mathbf{w}_w$,$\mathbf{w}_h$), the predicted bounding box of proposal (P) can be calculated as follows:

$$\hat{P}_x = P_w \mathbf{w}_x{}^T \phi_5(P) + P_x, \quad (6)$$

$$\hat{P}_y = P_h \mathbf{w}_y{}^T \phi_5(P) + P_y, \quad (7)$$

$$\hat{P}_w = P_w \exp(\mathbf{w}_w{}^T \phi_5(P)), \quad (8)$$

$$\hat{P}_h = P_h \exp(\mathbf{w}_h{}^T \phi_5(P)). \quad (9)$$

The new predicted proposals have more accurate location accuracy.

Though RCNN dramatically improves the object detection performance, the object proposals should be warped into a fixed size and then put into the CNN network, respectively. Because the computation of CNN features of different proposals are not shared, RCNN is very time-consuming. To remove the fixed-size constraint and accelerate detection speed, He et al. [25] proposed SPPnet. Figure 4 compares RCNN and SPPnet. Instead of cropping or warping the image data of all the proposals before computing the CNN features, SPPnet firstly computes all the convolutional features of the whole image and then uses spatial pyramid pooling to extract the fixed-size features of each proposal. Figure 5 gives the illustrationof spatial pyramid pooling layer (SPP). Based on the feature maps of last convolutional

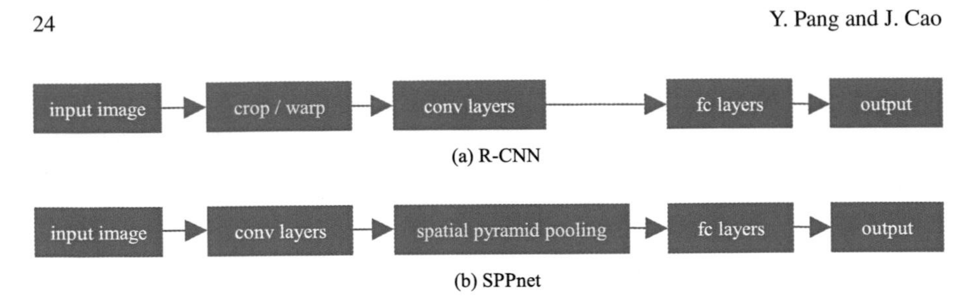

Fig. 4 Comparison of RCNN and SPPnet

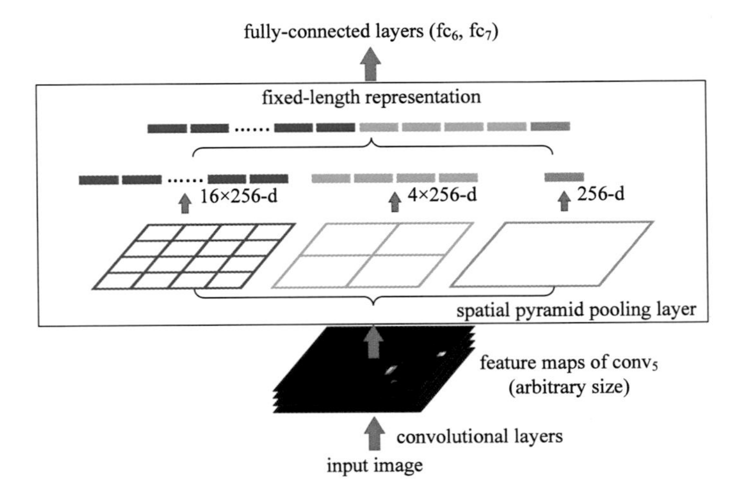

Fig. 5 Spatial pyramid pooling layer. The feature maps of a given proposal are pooled into 3×3 spatial bins, 2×2 spatial bins, and 1×1 spatial bins, respectively. After that, they are concatenated into a fixed-size feature vector and fed into two fully connected layers

layer, SPP splits the feature maps of the proposal into 3×3 spatial bins, 2×2 spatial bins, and 1×1 spatial bins, respectively. In each spatial bin, the feature response value is calculated as the maximum of all the features which belong to the same spatial bin (i.e., max-pooling). After that, the outputs of 3×3 spatial bins, 2×2 spatial bins, and 1×1 spatial bins are concatenated as a $21c$-d feature vector, where $c = 256$ is the number of feature maps of the last convolutional layer. After concatenation, two fully connected layers are connected to this $21c$-d feature vector. For training the CNN network of SPPnet, two different strategies can be adopted: single-scale training and multi-scale training. Single-scale training uses a fixed-size input (i.e., 224×224) wrapped from the input images. Multi-scale training uses the images of multiple different sizes, where in each iteration, only the images of one scale are used for training the CNN network. The size of input image is represented by $s \times s$, where s is uniformly sampled from 180 to 224. Because the multi-scale training can simulate the varying sizes of images, it can improve detection accuracy. For training SVM classifiers, the specific steps are the same as that of RCNN. In the test stage, the image of arbitrary scale can be put into SPPnet. Compared to

RCNN, SPPnet has the following advantages: (1) All the candidate proposals share all the convolutional layers before the fully connected layers. Thus, it has faster detection speed than RCNN. (2) SPPnet makes use of multilevel spatial information of objects, which is more robust to object deformations. Meanwhile, multi-scale training can enlarge the training data. Thus, it has the higher detection accuracy than RCNN.

Though SPPnet accelerates detection speed by sharing computation of all the convolutional layers, the training of SPPnet is still a multistage process similar to RCNN. Namely, they need to, respectively, fine-tune the CNN network on object detection dataset, train multiple SVM classifiers, and learn the bounding box regressors. To fix the disadvantages of RCNN and SPPnet, Girshick [24] further proposed Fast RCNN, which integrates the training of CNN networks, object classification, and bounding regression into a unified framework. Figure 6 shows the architecture of Fast RCNN. Generally, Fast RCNN firstly calculates all the convolutional layers of the whole image. For each proposal, Fast RCNN uses a ROI pooling layer to extract the fixed-size feature maps from the feature maps of last convolutional layer, then feds the fixed-size feature maps to two fully connected layers, and finally generates two sibling branches with fully connected operation for object classification and box regression. For object classification, it has $c+1$ outputs by softmax, where c means the number of object classes. For box regression, it has $4c$ outputs, where each four outputs correspond to the box offset per class. The ROI pooling layer warps the feature maps of object proposals into the fixed-size spatial bins (e.g., 7×7) and uses max-pooling operation to calculate the feature responses in each bin. Because Fast RCNN has two sibling outputs for object classification and box regression, the multitask training loss (i.e., L) is the joint of classification loss (i.e., L_{cls}) and regression loss (i.e., L_{loc}) for each ROI as follows:

$$L(p, t) = L_{cls}(p, c) + \lambda[c \geq 1]L_{reg}(t, v), \tag{10}$$

where λ balances classification loss and regression loss and $[c \geq 1]$ is equal to 1 if the $c \geq 1$ and 0 otherwise. Namely, the ROI belonging to background class does

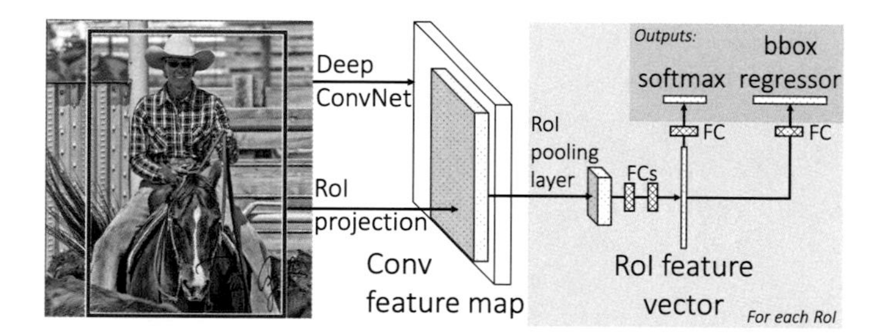

Fig. 6 The architecture of Fast RCNN. Compared to RCNN and SPPnet, it joins the classification and regression into a unified framework. (©[2015] IEEE. Reprinted, with permission, from Ref. [24])

not contribute to the regression loss. The classification loss $L_{cls}(p, c) = -\log p_c$ is the log loss for true class c. The regression loss L_{reg} is defined by the ground-truth regress target (i.e., (v_x, v_y, v_w, v_h)) and a predicted target (i.e., (u_x, u_y, u_w, u_h)) as follows:

$$L_{loc}(t, v) = \sum_{i \in x,y,w,h} smooth_{L_1}(t_i, v_i), \tag{11}$$

where

$$smooth_{L_1}(x) = \begin{cases} 0.5x^2, & \text{if } |x| < 1, \\ |x| - 0.5, & \text{otherwise.} \end{cases} \tag{12}$$

Compared to L_2 loss used in RCNN and SPPnet, the L_1 loss is more robust to outliers.

In the training stage, the CNN network is firstly initialized from the pre-trained ImageNet network and then trained with back-propagation in the end-to-end way. The mini-batches (i.e., R) are sampled from N images, where each image provides the R/N proposals. The proposals from the same image share all the convolutional computation. Generally, N is set as 2 and R is set as 128. To achieve scale invariance, two different strategies are used: brute-force approach and multi-scale approach. In brute-force approach, the images are resized to a fixed size in the training and test stages. In the multi-scale approach, the images are randomly rescaled to a pyramid scale in each iteration of training stage. In the test stage, the multi-scale images are put into the trained network, respectively, and the detection results are combined together by NMS.

For proposal generation, RCNN [23], SPPnet [25], and Fast RCNN [24] are all based on selective search. Selective search [60] uses the handcrafted features and adopts the hierarchical grouping strategies to capture all possible object proposals. Generally, it runs at 2 s per image on the common CPU. The detection network of Fast RCNN can run at about 100 ms per image on the GPU. Thus, proposal generation of Fast RCNN is more time-consuming compared to the detection network of Fast RCNN. Though selective search can be also re-implemented on the GPU, proposal extraction is still isolated from detection network of Fast RCNN. Thus, region proposal extraction becomes the bottleneck of Fast RCNN on object detection. To solve this problem, Ren et al. [52] proposed Faster RCNN. It integrates proposal generation, proposal classification, and proposal regression into a unified network. Figure 7 shows the network architecture of Faster RCNN. It consists of two modules. One module is called region proposal network (i.e., RPN), which is used to extract candidate object proposals. Another module is Fast RCNN, which is used to classify these proposals into the specific categories and predict more accurate proposal locations. The two modules share the same base sub-network. On the one hand, RPN can generate the candidate object proposals by deep convolutional features; it can improve proposal location quality. On the other hand, Faster RCNN

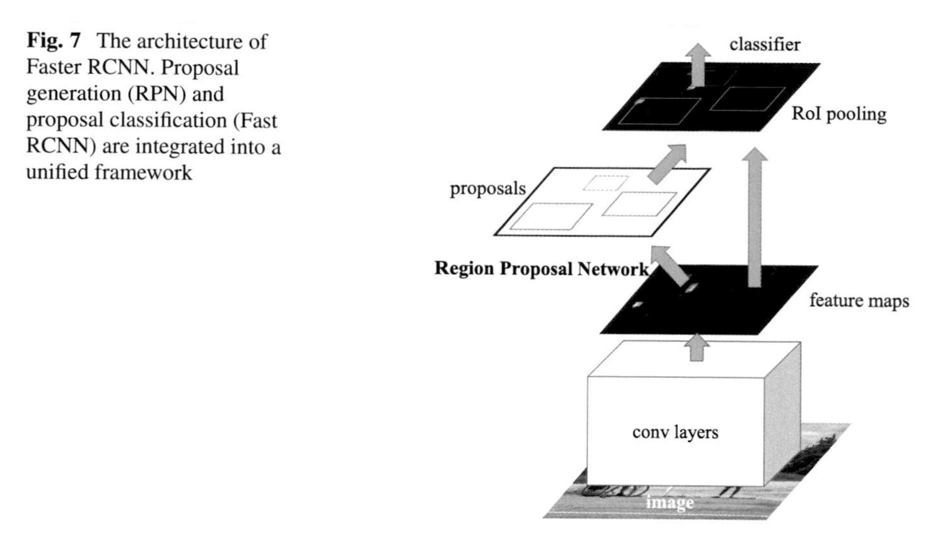

Fig. 7 The architecture of Faster RCNN. Proposal generation (RPN) and proposal classification (Fast RCNN) are integrated into a unified framework

is an end-to-end framework with the multitask loss. Compared to Fast RCNN, Faster RCNN can achieve better detection performance with much fewer proposals.

RPN slides a small network over the output layer of base network. The small network consists of one 3×3 convolutional layer and two sibling 1×1 convolutional layers for box regression and box classification. Box classification is class-agnostic. For each sliding window location, RPN predicts multiple proposals based on the anchors of different aspect ratios and scales. Assuming that the number of anchors is k, the box regression layer has $4k$ outputs for each sliding window, and the box classification layer has $2k$ outputs for each sliding window. Generally speaking, three different aspect ratios of $\{1 : 2, \ 1 : 1, \ 2 : 1\}$ and three different scales of $\{0.5, 1, 2\}$ are used. Thus, there are nine (i.e., 3×3) anchors (i.e., $k = 9$) at each sliding window. The multitask loss of RPN consists of two parts: classification loss L_{cls} and regression loss L_{reg}, which can be written as follows:

$$L(p, v) = \frac{1}{N_{cls}} \sum_i L_{cls}(p_i, c_i) + \lambda \frac{1}{N_{reg}} \sum_i [c_i \geq 1] L_{reg}(t_i, v_i) \tag{13}$$

where N_{cls} (256) and N_{reg} (about 2400) are the terms to, respectively, normalize classification loss and location loss, λ balances classification loss and regression loss, and $[c_i \geq 1]$ is 1 if $c_i \geq 1$ or 0 otherwise. The classification loss and regression loss are the same as that of Fast RCNN. Two following kinds of anchors are labelled as the positives: (1) the anchor with the highest IoU overlap with a ground-truth bounding box and (2) the anchor that has an IoU overlap over 0.7 with any ground-truth bounding boxes. The anchors are labelled as the negatives if the anchors are not labelled as the positives and they have the IoU overlap under 0.3 with all ground-truth bounding boxes.

Because RPN and Fast RCNN share the same base network, thus they cannot train independently. Faster RCNN provides three different ways to train RPN and Fast RCNN in the unified network: (1) Alternating training. RPN is firstly trained. Based on the proposals generated by RPN and the filter weights of RPN, Fast RCNN is then trained. Two above steps are iterated for two times. (2) Approximate joint training. RPN and Fast RCNN networks are seen as a unified network. The loss of the unified network is the joint of RPN loss and Fast RCNN loss. In each iteration of training stage, the proposals generated by RPN are treated as the fixed proposals when training Fast RCNN detector. Namely, the derivative of proposals coordinates are ignored. (3) Non-approximate joint training. The difference between approximate joint training and non-approximate joint training is that the derivative of proposal coordinates should be considered. Because the standard ROI pooling layers are not differentiable for proposal coordinate, the first two ways are usually used.

Generally, the above state-of-the-art object detection methods (i.e., RCNN [23], SPPnet [25], Fast RCNN [24], and Faster RCNN [52]) use the pre-trained CNN network on image classification dataset (i.e., ImageNet [13]). Dai et al. [11] argued that this design has the dilemma in some degree. Generally, deep CNN network for image classification usually favors translation invariance, while deep CNN network for object detection needs to be translation variance. To address the above dilemma between image classification and object detection, Dai et al. [11] proposed R-FCN. It encodes the object position information by the position-sensitive ROI pooling layer (PSROI) for the following Fast RCNN subnet. Figure 8 shows the architecture of R-FCN. Region proposal generation is the same as Faster RCNN. Based on the output layer of original base network, R-FCN generates the new $k * k$ position-sensitive convolutional banks. The convolutional banks correspond to the $k \times k$ spatial grids, respectively. In each convolutional bank, there are $c + 1$ convolutional layers (c means the number of object categories, and $+1$ means the background category). Namely, $k * k * (c + 1)$ convolutional feature maps are generated. For

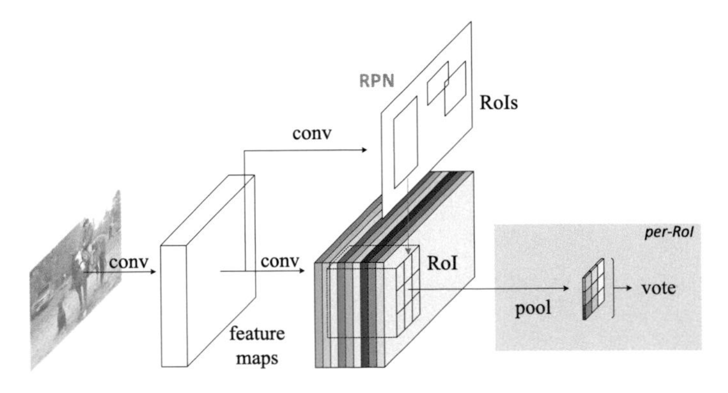

Fig. 8 The architecture of R-FCN. Position information is encoded into the network by position-sensitive ROI pooling (PSROI)

a given proposal, position-sensitive ROI pooling layer generates the $k \times k$ score maps from position-sensitive feature maps, where the response of (i, j)-th bin (i.e., $r(i, j)$) pools over (i, j)-th positive-sensitive score maps as follows:

$$r_{r^*}(i, j) = \sum_{(x,y)\in bin(i,j)} z_{i,j,c^*}(x + x_0, y + y_0)/n, \quad (14)$$

where z_{i,j,c^*} is one of $k * k * (c + 1)$ feature maps, (x_0, y_0) denotes top-left corner of a ROI proposal, n is the number of pixels in the $bin(i, j)$, $r_{c^*}(i, j)$ is the pooled response of $bin(i, j)$ for c^*-th category. After that, average pooling or max-pooling is conducted to output the scores for object classification. For box regression, sibling $4 * k * k$ convolutional feature maps are also generated, and then position-sensitive ROI pooling operation is conducted to construct four feature maps with the size of $k \times k$. Finally, 4-d vector of the box parameter is calculated by the average voting. For example, if k is set as 3, it means that the position information is encoded as $\{top - left, top - center, \ldots, bottom - right\}$.

The loss of R-FCN is similar to that of Faster RCNN. Please note that the proposal location predications per category in Faster RCNN are based on different outputs of box regression layer, while all the proposal location predications in R-FCN share the same output of box regression layer. R-FCN is fully connected networks, which almost shares all the CNN computation of the whole image. Thus, it can achieve the competitive detection accuracy and faster detection speed compared to Faster RCNN.

Most methods of object detection only predict object locations by bounding box and do not provide the more accurate segmentation information. In recent few years, some researchers proposed instance segmentation, which usually contains object detection and segmentation. Mask RCNN is a famous method for instance segmentation and object detection. Figure 9 shows the architecture of Mask RCNN. Mask RCNN incorporates instance segmentation and object detection into a unified framework based on Faster RCNN architecture. Specifically, it adds an extra mask

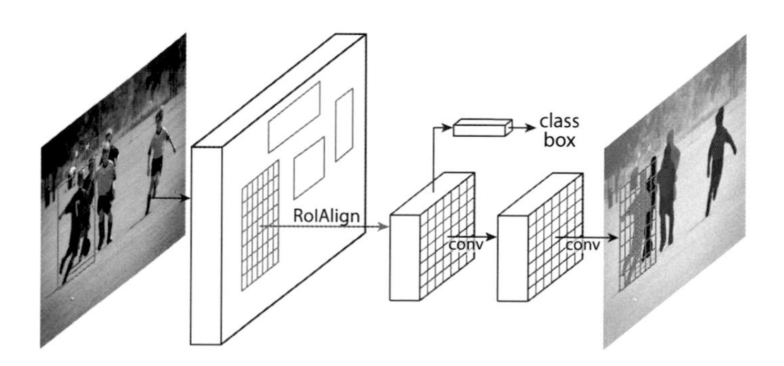

Fig. 9 The architecture of Mask RCNN. Apart from detection branch of Faster RCNN, the extra branch of mask segmentation is added. (©[2017] IEEE. Reprinted, with permission, from Ref. [27])

branch to predict the mask of object aside from the branch for object classification and box regression. The mask branch has c binary masks with the size of $m \times m$. c means the number of object categories. The multitask loss of Mask RCNN on each sampled ROI is the joint of classification loss, regression loss, and mask loss. It can represented as $L = L_{cls} + L_{reg} + L_{mask}$. The losses of L_{cls} and L_{reg} are the same as that of Faster RCNN. For an RoI proposal associated with the ground-truth class c^*, L_{mask} is only defined on the c^*-th mask, and other mask outputs do not contribute to the loss. Based on this design, it allows the network to generate masks for every class without competition among classes. In test stage, the output mask of object is determined by the predicted category of classification branch. To extract a small feature map for each ROI, ROI pooling quantizes the floating number of ROI proposal location into the discrete values. The quantization of ROI pool causes the misalignment between the input and the output, which has a negative effect on instance segmentation. To solve this problem, ROIAlign is proposed. It uses bilinear interpolation to compute the feature values of four corner locations in each spatial bin and then aggregates the feature response of each bin by max-pooling. Based on multitask learning, Mask RCNN can achieve state-of-the-art performance on object detection and instance segmentation. It means that joining the instance segmentation task with object detection task can also help improve detection performance.

Some other improvements of two-stage methods are also proposed. Yang et al. [67] proposed a "divide and conquer" solution-based Fast RCNN, which firstly rejects the background from the candidate proposals and then uses the c trained category-specific network to judge the proposal category. Based on the classification loss of proposals, Shrivastava et al. [55] proposed online hard example mining to select the hard samples. Gidaris et al. [22] proposed LocNet to improve location accuracy, which localizes the bounding box by the probabilities on each row and column of a given region. Bell et al. [1] proposed to use the contextual information and skip-layer pooling to improve detection performance. Kong et al. [31] proposed to use skip-layer connection for object detection. Zagoruyko et al. [70] proposed to use multiple branches to extract object context of multiple resolutions. Yu et al. [69] proposed dilated residual networks to enlarge the resolution of output layer without reducing the respective field.

2.2 *One-Stage Methods for Deep Object Detection*

Different from the multistage process of two-stage methods, one-stage methods aim to simultaneously predict object category and object location. Compared to two-stage methods, one-stage methods have much faster detection speed and comparable detection accuracy. Among the one-stage methods, OverFeat [54], YOLO [49], SSD [38], and RetinaNet [36] are the representative methods.

YOLO [49] divides the input image into the $k \times k$ grids. Each grid cell predicts B bounding boxes with objectness scores and c conditional class probabilities. The predictions of each bounding box are (x, y, w, h, s), where (x, y, w, h) gives the location of bounding box and s is the confidence objectness score of bounding box. Thus, the output layer has the size of $n \times n \times (5B + c)$. Based on bounding box prediction and corresponding class prediction, YOLO can simultaneously give the object probability and the category probability of each cell. Figure 10 shows the architecture of YOLO. It consists of 24 convolutional layers and 2 fully connected layers. To accelerate detection speed, the alternating 1×1 convolutional layers are used in some middle layers. The input image size of YOLO is fixed (i.e., 448×448). On PASCAL VOC, B is set to 2, and c is 20. Thus, the output layer for PASCAL VOC has the size of $7 \times 7 \times 30$. The loss of YOLO is the joint of classification loss, location loss, and detection loss.

In the training stage, the first 20 convolutional layers of YOLO are firstly pre-trained, and the whole network of YOLO is then fine-tuned on the dataset of object detection. YOLO uses sum-squared error for optimization. Because there are many grid cells that do not contain objects, it will affect the gradient from the cells that contain objects. Thus, the weight of box location predication loss is increased, while the weight of box predication loss is decreased. For each bounding box, it is assigned to the ground-truth with which it has the highest IoU overlap.

SSD [38] is a single-shot detector for generic object detection. The base network is based on VGG16 [56]. Following the base network, several convolutional layers are added to generate more convolutional layers (i.e., C6-C11) of different resolution. After that, it uses multiple convolutional layers of different resolution to predict objects of different scales. Specifically, for the output convolutional layer of a given resolution, it firstly uses a 3×3 convolutional filter to generate the new feature maps and then predicts object category scores and object locations on the new feature maps. Figure 11 shows the architecture of SSD. Assuming that the number of object classes is c and each feature map predicts k objects, it will result

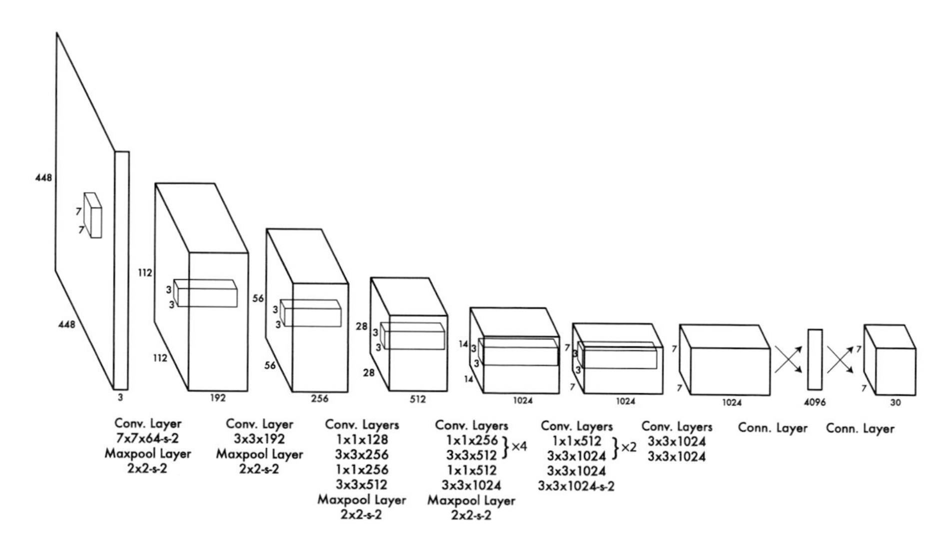

Fig. 10 The architecture of YOLO. (©[2016] IEEE. Reprinted, with permission, from Ref. [49])

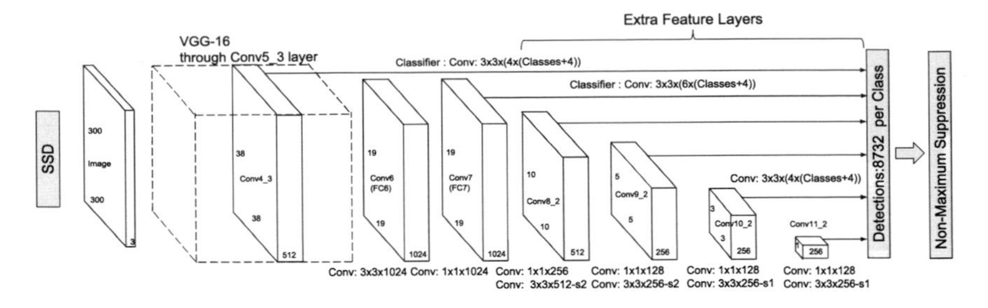

Fig. 11 The architecture of SSD. The base network is VGG16. (Reprinted from Ref. [38], with permission of Springer)

$(c + 4) * k * m * n$ output vector for the given $m \times n$ feature maps. The objective loss of SSD is a weighted sum of location loss and confidence loss similar to that of Fast RCNN.

For multi-scale object detection based on different convolutional layers, the anchor scale for each convolutional layer is computed as:

$$s_k = s_{min} + \frac{s_{max} - s_{min}}{K - 1}(k - 1),\ k \in [1, K], \tag{15}$$

where s_{min} and s_{max} are 0.2 and 0.9 and K means the number of different convolutional layers used for prediction. Aspect ratios of anchors are set as $\{1, 2, 3, 1/2, 1/3\}$. For the aspect ratio of 1, the anchor with extra scale of $\sqrt{s_k * s_{k+1}}$ is also added. Thus, there are six default boxes per feature map location. The default anchor is labelled as the positive for the matched class if it has the highest Jaccard overlap with a ground-truth or it has the Jaccard overlap over 0.5 with the ground-truth. Namely, SSD can predict high scores for multiple overlapping anchors. Generally, the negative and positive training samples have a significant imbalance. The bootstrap technique is used. According to the highest confidence loss, the anchors of negatives are sorted. Then, some top negatives are used so that the ratio of positives and negatives is about 1:3. Compared to YOLO, SSD makes full use of multi-scale information. Thus, it has a faster detection speed and higher detection accuracy.

Though one-stage methods (e.g., YOLO and SSD) have the faster detection speed than two-stage methods, most state-of-the-art methods for object detection are still two-stage. Lin et al. [36] investigated why one-stage methods have the inferior detection performance compared to the two-stage methods. It was found that the extreme positive and negative imbalance is the main reason that causes the inferior performance of one-stage methods. To solve the imbalance of positives and negatives, bootstrap technique is usually used to choose some hard negatives. However, it will ignore the information of easy negative. If all the negatives are used, the weights of easy negatives will be large which causes the worse detection performance. To solve this problem, RetinaNet is proposed. RetinaNet adopts the focal loss for object detection based on FPN architecture. Focal loss can be also seen as a bootstrap technique. In the training stage, it (i.e., $FL(p_t)$) adds a factor to the cross entropy loss as follows:

$$FL(p_t) = -(1 - p_t)^{\gamma} \log(p_t) \tag{16}$$

where

$$p_t = \begin{cases} p, & \text{if } c = 1, \\ 1 - p, & \text{otherwise,} \end{cases} \tag{17}$$

where p is the probability of the anchor which belongs to class $c = 1$ and $\gamma > 0$. Based on this operation, it can enlarge the weights of hard negative samples and reduce the weights of easy samples. Namely, it can pay more attention on the hard negatives and reduce the influence of easy negatives. Moreover, α-balanced variant of focal loss is further proposed as follows:

$$FL(p_t) = -\alpha(1 - p_t)^{\gamma} \log(p_t), \tag{18}$$

where α belongs to [0,1]. With the α-balanced variant of focal loss, it can further improve detection accuracy.

Figure 12 shows the architecture of RetinaNet. The base network is FPN, which constructs multiple feature maps. For box classification and box regression, four 3×3 convolutional layers are attached to the output layer of FPN, respectively. After that, 3×3 convolutional layers with $K * C$ filters are used for box classification, and 3×3 convolutional layers with $4 * K$ filters are used for box regression. Namely, one position of each layer has K anchors, and there are C classes. The anchors are assigned to the matched class if it has the highest IoU overlap with a ground-truth, and the IoU overlap is over 0.5. The anchor is assigned to background if it has the highest IoU overlap below 0.4 with a ground-truth. The other anchors are ignored in the training stage. The anchors have areas of 32×32 to 512×512 on pyramid levels P3 to P7. The aspect ratios for each feature map are set as $\{1/2, 1, 2\}$, respectively. Focal loss can make full use of all the information of negatives in the training stage, while the easy negatives have the relatively small influence. Compared to SSD, FPN

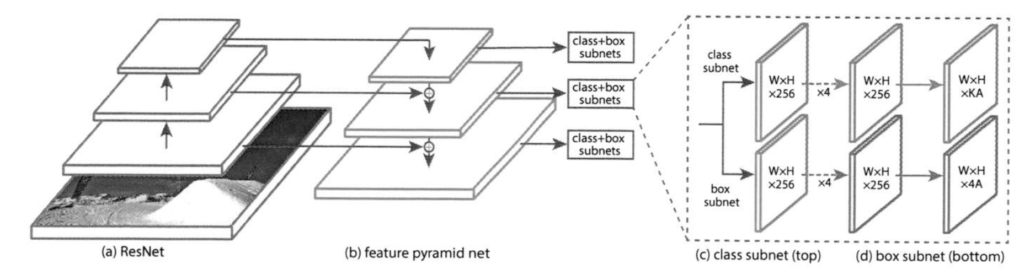

Fig. 12 The architecture of RetinaNet. The base network is FPN with ResNet. (©[2017] IEEE. Reprinted, with permission, from Ref. [36])

uses the top-down structure to enhance the semantic levels of high-resolution and low-level semantic convolutional layers.

Some other one-stage methods are also proposed. Najibi et al. [42] proposed to initially divide the input image into the multi-scale regular grids and iteratively update the location of these grids to be toward the objects. Based on SSD, Ren et al. [50] further proposed to use the recurrent rolling convolution to add deep context information. Fu et al. [20] proposed to use deconvolutional layer after SSD to incorporate the object context.

3 Pedestrian Detection

As the canonical and important case of object detection, pedestrian detection can be applied into many areas (autonomous driving, human-computer interaction, video surveillance, and robotics). In the past 10 years, pedestrian detection has also achieved great success [3]. Figure 13 shows the progress of pedestrian detection Caltech pedestrian dataset. According to whether using CNN, pedestrian detection can be mainly classified into two classes: handcrafted feature-based methods and CNN-based methods.

3.1 Handcrafted Feature-Based Methods for Pedestrian Detection

In 2004, Viola and Jones [61] proposed robust real-time face detection, which uses cascaded AdaBoost to learn strong classifier from the candidate Haar feature pool. In 2005, Dalal and Triggs [10] proposed histograms of gradient (HOG) for

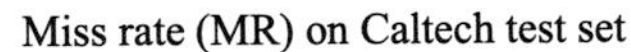

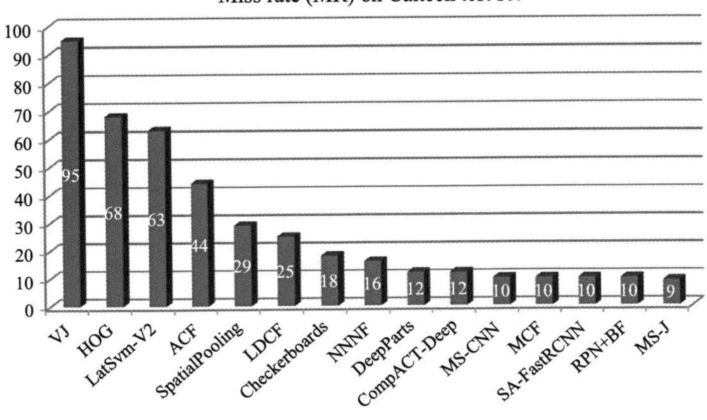

Fig. 13 The progress of pedestrian detection on Caltech pedestrian dataset

pedestrian detection, which dramatically improves the performance of pedestrian detection. Inspired by the above two ideas, Dollár et al. [14] integrated them into pedestrian detection, which is called integral channel features (i.e., ICF). The top row of Fig. 14 shows the architecture of ICF. It firstly converts the original input image into ten feature channels (i.e., HOG+LUV), then extracts the features of local pixel sum in each channel, and finally learns the strong classifier from the candidate feature pools by cascaded AdaBoost. HOG+LUV consists of six histograms of gradient, one gradient magnitude, and the LUV color channels. The features of local pixel sum are efficiently calculated by using integral images. To accelerate detection speed, Dollár et al. [16] further proposed aggregated channel features (i.e., ACF). It downsamples the original feature channels (i.e., HOG+LUV) four times and uses every single pixel value in each channel as the candidate feature. After that, some variants of ICF are also proposed. To avoid the randomness of candidate features in ICF, SquaresChnFtrs [2] deterministically generates candidate features by calculating the pixel sum features of all the squares inside each channel. To reduce the local correlation in each feature channel, LDCF [43] convolves the filters learned by PCA technique with feature channels (i.e., HOG+LUV) to generate new feature maps.

By observing and summarizing ICF and the variants of ICF, Zhang et al. [73] generalized these methods into a unified framework. It is called filtered channel features (FCF). The bottom row of Fig. 14 shows the pipeline of filtered channel features (FCF). It firstly converts the input image to the HOG+LUV channels, then convolves the filter bank with HOG+LUV to generate the new feature channels

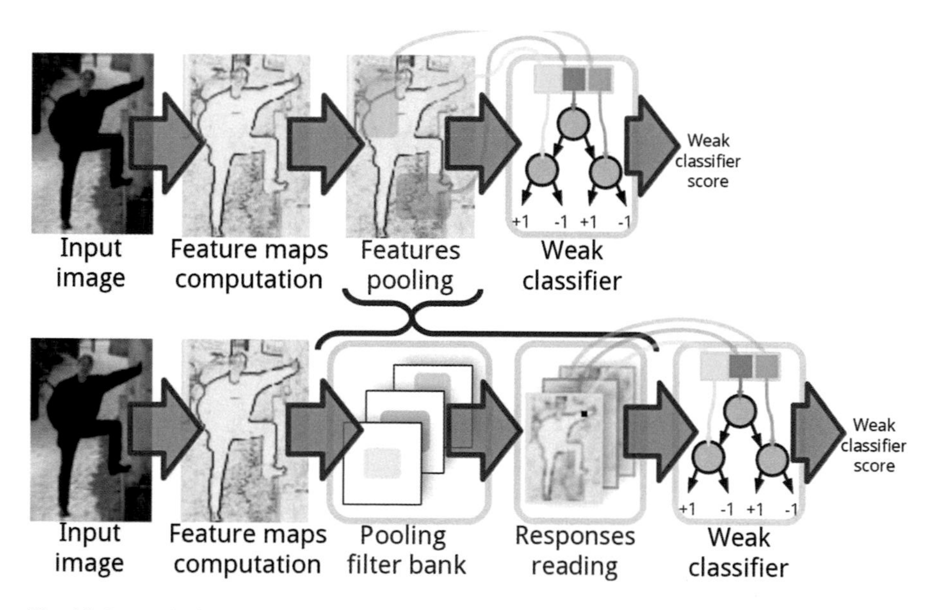

Fig. 14 Integral channel features (ICF) and filtered channel features (FCF). (©[2015] IEEE. Reprinted, with permission, from Ref. [73])

where each pixel value in each channel is set as candidate feature, and finally learns the strong classifier by decision forest from the candidate feature pool. The filter bank can be different types, including SquaresChntrs filters, Checkerboards filters, Random filters, Informed filters, and so on. Thus, ICF [14], ACF [16], SquaresChntrs [2], and LDCF [43] can be seen as the special cases of FCF. It is found that based on Checkerboards filters, FCF can outperform ICF, ACF, SquaresChntrs, and LDCF on Caltech pedestrian dataset [15] and KITTI benchmark [21].

These above methods make great success on pedestrian detection. However, the design of handcrafted features does not consider pedestrian inherent attributes. Zhang et al. [72] treated the pedestrians as three parts (i.e., head, upper body, and legs) and designed the Haar-like features (i.e., InformedHaar) for pedestrian detection. However, it still does not make full use of pedestrian attributes. To make better use of pedestrian attributes for pedestrian detection, Cao et al. [6, 7] further proposed two non-neighborhood features inspired by appearance constancy and shape symmetry of pedestrians.

Generally, pedestrian can be seen as three parts: head, upper body, and legs. Usually, the appearance of these parts is constancy and contrast to the surrounding background. Based on appearance constancy, side-inner difference features (SIDF) are proposed. Figure 15 gives the illustration of side-inner difference features (SIDF). Patch B is randomly sampled between patch A and the symmetrical patch A'. The height of patch B is the same as that of patch A, while the width of patch B can be different from patch B. The direction of patch A can be horizontal or vertical, while the sizes of patch A can be arbitrarily various in the maximum square of 8×8 cells, where each cell is 2×2 pixels. SIDF (i.e., $f(A, B)$) calculates the difference between the local patch (i.e., patch A) in background and the local patch (i.e., patch B) in pedestrians as:

$$f(A, B) = \frac{S_A}{N_A} - \frac{S_B}{N_B}, \tag{19}$$

Fig. 15 Illustration of side-inner difference features (SIDF). SIDF calculates the difference of patch A and patch B, where patch B is randomly sampled between patch A and the symmetrical patch A'

where S_A and S_B are, respectively, the pixel sums in patch A and patch B and N_A and N_B means, respectively, the pixel number in patch A and patch B.

Meanwhile, pedestrians usually appear stand-up. Thus, the shape of pedestrian is loosely symmetrical in the horizontal direction. Based on shape symmetry, symmetrical similarity features (SSF) are proposed. Figure 16 shows symmetrical similarity features (SSF). Patch A and Patch A' are two symmetrical patches on pedestrians. The direction of patch A can be horizontal or vertical, and the size of patch A can be changed from 6×6 cells to 12×12 cells, where each cell is 2×2 pixels. SSF (i.e., $f(A, A')$) calculates the similarity between two symmetrical patches (i.e., patch A and patch A') of pedestrians as follows:

$$f(A, A') = |f_A - f_{A'}| = |\frac{S_A}{N_A} - \frac{S_{A'}}{N_{A'}}|, \tag{20}$$

where f_A and $f_{A'}$ represent the features of patches A and A'. Because the shape symmetry of pedestrians is not very strict, max-pooling technique is further incorporated to improve the robustness of the features. Specifically, the feature value (i.e., $f_M(A)$) of patch A is represented by the maximum feature of three sub-patches in patch A (i.e., patch A_1, patch A_2, and patch A_3) as follows:

$$f_M(A) = \max_{i=1,2,3} \frac{S_i}{N_i}. \tag{21}$$

The three sub-patches are randomly sampled in patch A or patch B. Thus, they can have different aspect ratios, positions, and sizes. Based on it, SSF (i.e., $f(A, A')$) can be rewritten by:

$$f(A, A') = |f_M(A) - f_M(A')|. \tag{22}$$

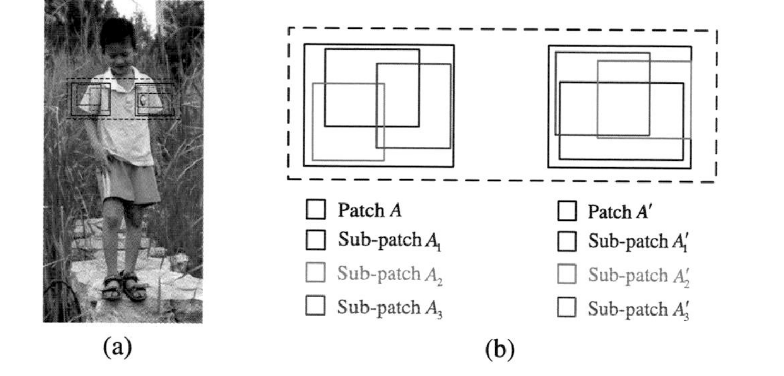

Fig. 16 Illustration of symmetrical similarity features (SSF). SSF abstracts the similarity between patch A and the symmetrical patch A'. Each patch is represented by three random sub-patches

Because SIDF and SSF are calculated by non-neighboring patches, they are called non-neighboring features (NNF). To achieve state-of-the-art detection performance, neighboring features (NF) are also designed for pedestrian detection. Figure 17 shows the neighboring features, which contain local mean features and neighboring difference features. In Fig. 17a, the sizes and aspect ratios of local mean features can be changed. In Fig. 17b, partition line is where two neighboring patches intersect. The direction of partition line can be horizontal or vertical. The location of partition line can be also various.

Based on Caltech2x training set, the detectors which select features from NF, NF+SIDF, NF+SSF, or NNNF are trained. Level-2 2048 decision forests are used. Table 1 compares NF, NF+SIDF, NF+SSF, and NNNF. It can be seen with non-neighboring features (SIDF or SSF); NF+SIDF or NF+SSF can outperform NF by 1.83% or 2.30%. When SIDF and SSF are both combined with NF, NNNF outperforms NF by 4.44%. Figure 18 shows the ratios of NF, SSF, and SIDF in NNNF. It can be seen that 69.97% features are NF, 11.34% features are SSF, and 18.69% features are SIDF. Namely, about 70% features are neighboring features, and about 30% features are non-neighboring features. It demonstrates the

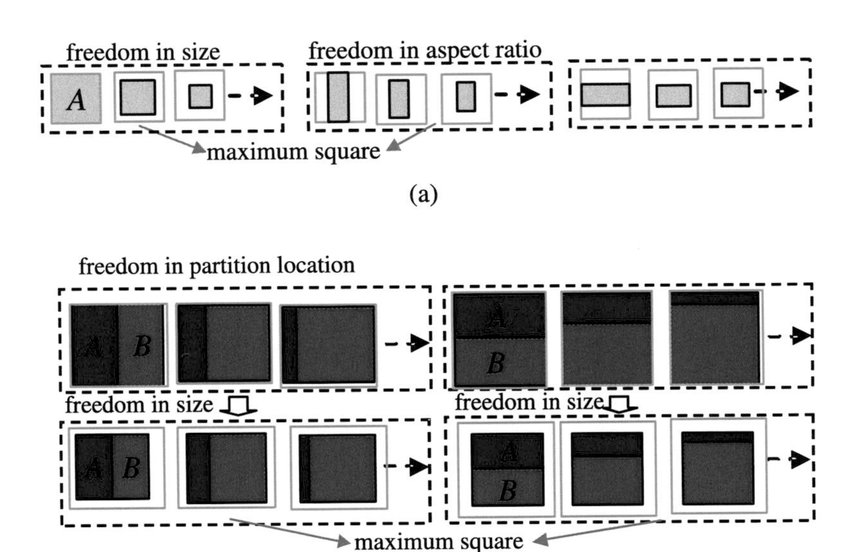

Fig. 17 Illustration of neighboring features. (**a**) is the local mean feature, and (**b**) is the neighboring feature

Table 1 Comparison of log-average miss rates on Caltech test set

Method	MR	Δ MR
NF	27.50%	N/A
NF+SIDF	25.67%	+1.83%
NF+SSF	25.20%	+2.30%
NNNF	23.06%	+4.44%

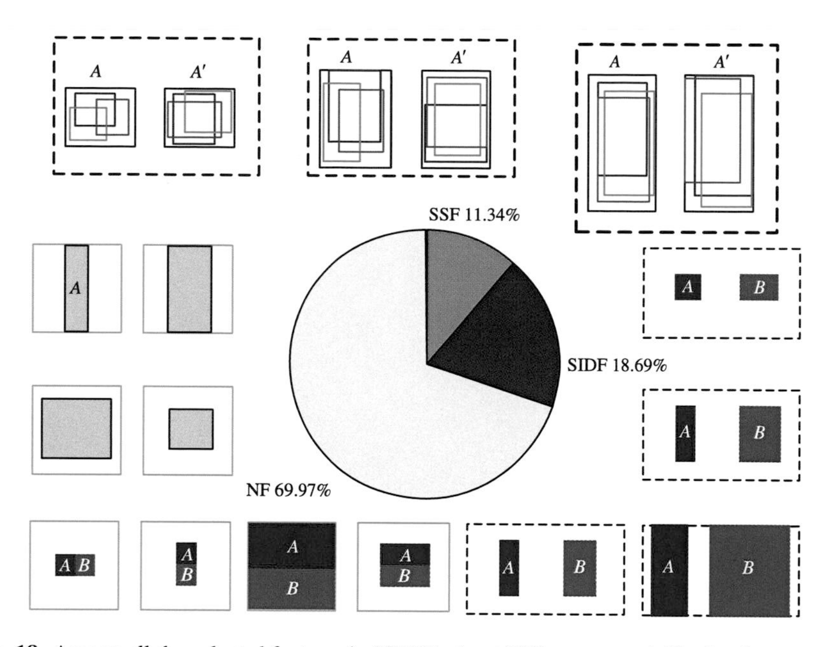

Fig. 18 Among all the selected features in NNNF, about 30% are non-neighboring features, and 70% are neighboring features. Several representative non-neighboring features and neighboring features are also shown

Table 2 Average precision (AP) of some methods without using CNN on KITTI test set

Method	Easy	Moderate	Hard
ACF [16]	44.49%	39.81%	37.21%
SquaresChnFtrs [2]	57.33%	44.42%	40.08%
SpatialPooling+ [46]	65.26%	54.49%	48.60%
Checkerboards [73]	67.75%	56.75%	51.12%
NNNF-L4	**69.16%**	**58.01%**	**52.77%**

effectiveness of proposed non-neighboring features. Some representative features are also shown in Fig. 18.

To compare with the state-of-the-art methods, NNNF with 4096 level-4 decision forests are trained on Caltech10x training set. It is called NNNF-L4. Figure 19 shows ROC on Caltech test set. It can be seen that NNNF outperforms Checkerboards [73] by 2.27%. Table 2 shows the average precision (AP) on KITTI test set. It can be seen that NNNF outperforms Checkerboards by 1.36% on moderate test set. Namely, among the methods without using CNNN, NNNF achieves the state-of-the-art performance by combining NNF with neighboring features (NF) on Caltech pedestrian dataset and KITTI benchmark.

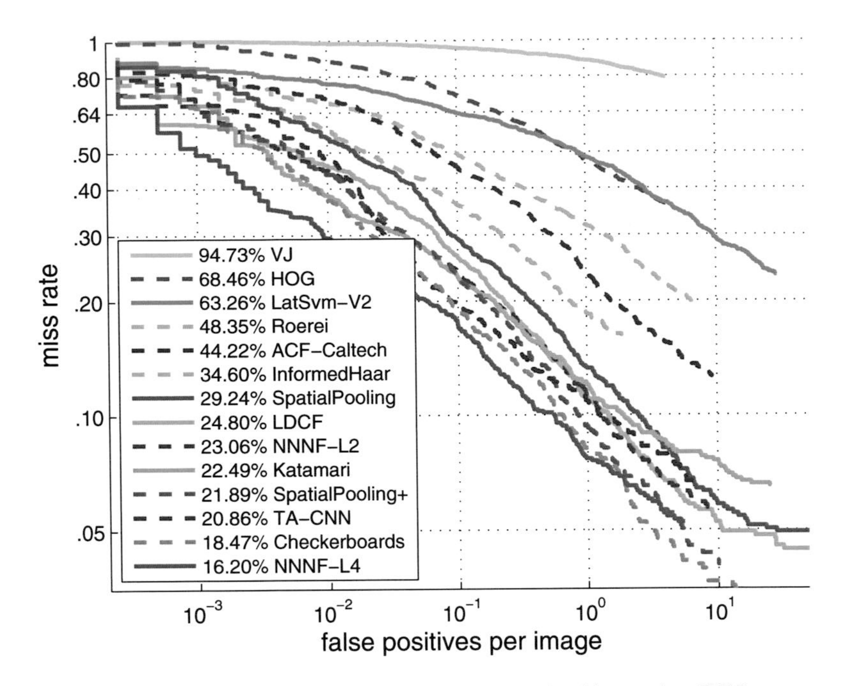

Fig. 19 ROC on Caltech test set (reasonable) of some methods without using CNN

3.2 CNN-Based Methods for Pedestrian Detection

With the success of convolutional neural networks on many fields of computer vision, researchers also explored how to apply CNN on pedestrian detection for improving detection performance. As an initial attempt, Hosang et al. [28] studied the effectiveness of CNN for pedestrian detection. In [28], it firstly extracts the candidate pedestrian proposals by using the handcrafted feature-based method (i.e., SquaresChnFtrs [2]) and then uses a small network (i.e., CifarNet) or a large network (i.e., AlexNet [33]) to classify these proposals. The traditional pedestrian detection usually uses a fixed-size detection window (i.e., 128×64). Following it, the input size of CNN network (i.e., CifarNet and AlexNet) is changed to be 128×64, and the outputs of CNN network are changed to be 2 (i.e., pedestrian or non-pedestrian). In the training stages, the proposal with an IoU overlap over 0.5 with a ground-truth bounding box is labelled as the positive, and the proposal with the IoU below 0.3 with all the ground-truth bounding boxes is labelled as the negative. The ratio of positives and negatives in the mini-batch is set as 1:5. Experimental results on Caltech pedestrian dataset demonstrate that the ConvNets can achieve the state-of-the-art performance, which is useful for helping pedestrian detection.

After that, pedestrian detection based on CNN achieves great success. Sermanet et al. [53] proposed to merge the down-sampled 1-st convolutional layer with 2-st convolutional layer to add the global information for pedestrian detection.

Instead of using handcrafted feature channel (HOG+LUV), Yang et al. [66] proposed convolutional channel features (CCF). CCF treats the feature maps of last convolutional layer as the feature channels and uses decision forests to learn the strong classifier. Because of the richer representation ability of CNN features, CCF achieves state-of-the-art performance on pedestrians detection. Zhang et al. [71] proposed to use RPN to extract candidate features and use decision forest to classify these proposals. Zhang et al. [74] proposed to improve the performance of Faster RCNN on pedestrian detection by some specific tricks (e.g., quantized RPN scales and the upsampled input image). To reduce the computation complexity, Cai et al. [4] proposed the complexity-aware cascaded detector (i.e., CompACT), which integrates the features of different complexities together. Complexity-aware cascaded detector aims to achieve a best trade-off between classification accuracy and computation complexity. In the training stage, the loss of CompACT is the joint of classification loss and computation complexity loss. Based on CompACT, the first few stages of strong classifier learn the more features of lower computation complexity, and the last few stages of strong classifier learn the more features of higher computation complexity. To further improve detection performance, the CNN features of very highest complexity are embedded into the last stage of CompACT, which is called CompACT-Deep. Because most detection proposals are rejected by the first few stages, only the CNN features of very few proposals need to be calculated. Thus, CompACT-Deep can improve detection performance without increasing much computation cost.

Some methods exploit to use the extra feature information to help pedestrian detection. Tian et al. [59] proposed to join pedestrian detection with semantic tasks, including scene attributes and pedestrian attributes, into the TA-CNN. The attributes are transferred from existing scene datasets. TA-CNN is learned by iterating two tasks, respectively. Mao et al. [41] proposed to apply some extra features to deep pedestrian detection. Some extra feature channels are used as the extra input channels for CNN detectors. The extra feature channels contain some low-level semantic feature channels (e.g., gradient and edge), some high-level sematic feature channels (e.g., segmentation and heatmap), depth channels, and temporal channels. It is found that segmentation and edge used as the extra input channels can significantly help improve pedestrian detection. Based on this observation, Mao et al. [41] further proposed HyperLearner. It consists of four different modules: base network, channel feature network (CFN), region proposal network, and Fast RCNN. Base network and feature, region proposal network, and Fast RCNN are the same as original Faster RCNN. Multiple convolutional layers of base network firstly go through two 3×3 convolutional layers and then upsample the same size of conv1. After that, the output layers are appended together to generate the aggregated feature maps. The aggregated feature maps are fed into CFN for channel feature prediction and concatenated with the output layer of base network. In the training stage, feature channel generation network is supervised by the semantic segmentation ground-truth. The loss of HyperLearner is the joint of Faster RCNN loss and segmentation loss. In the test stage, feature channel generation network outputs the predicted feature channel map (e.g., semantic segmentation map). Before Fast RCNN subnet,

ROI pooling layer pools over the concatenation of output layer of base network and aggregated features maps of CFN. With the help of extra feature information, HyperLearner improves detection performance.

Generally, CNN-based methods for pedestrian detection have the superior detection performance. However, the computation complexity of these methods is very high. Thus, these methods will be very slow when they run on the common CPU. Meanwhile, in many situations the computing device only has the CPU. Thus, speeding up CNN-based methods on CPU is very important and necessary. Compared to CNN-based methods, the traditional handcrafted feature-based methods are relatively simple and have the faster detection speed on the CPU. To accelerate detection speed on the common CPU, Cao et al. [8] proposed to integrate the traditional channel features and each layer of CNN into a multiple feature channels. Figure 20 shows the architecture of MCF. Firstly, multiple feature channels are constructed by HOG+LUV and each channel of CNN (e.g., AlexNet [33] and VGG16 [56]). Based on the multiple feature channels, the candidate features are further extracted. Finally, multistage cascade AdaBoost is learned from candidate features of corresponding layers. On the one hand, based on the handcrafted feature channels and each layer of CNN, MCF can learn more abundant features to improve detection performance. On the other hand, MCF can quickly reject many negative detection windows and thus reduce the computation of the remaining CNN layers. As a result, it can accelerate detection speed.

Table 3 shows the specific parameters of multilayer feature channels based on HOG+LUV and VGG16. It consists of six layers (i.e., L1, L2, ..., L6). L1 is the handcrafted feature channels (i.e., HOG+LUV). The size of HOG+LUV is 128×64. L2–L6 correspond to multiple convolutional layers of VGG16 (i.e., C1–C5). The sizes of L2–L6 are 64×32, 32×16, 16×8, 8×4, and 4×2, respectively. The numbers of channels in L1–L6 are 10, 64, 128, 256, 512, and 512, respectively. In fact, only part convolutional layers can be used to construct multilayer feature channels. For

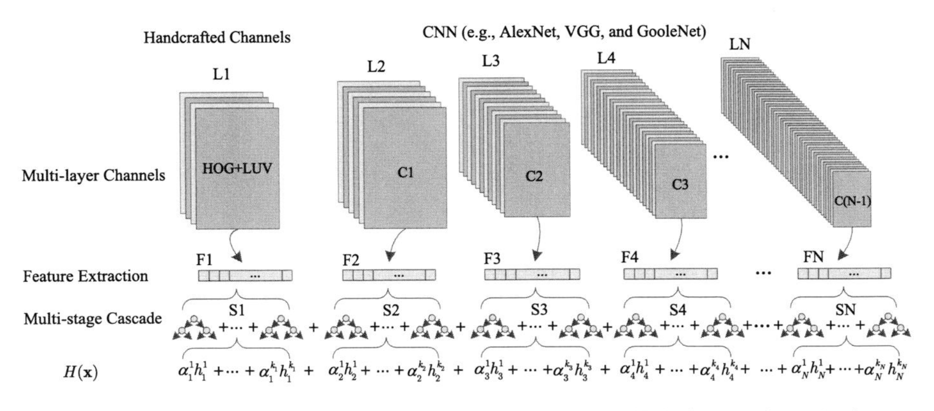

Fig. 20 The architecture of MCF. It consists of three parts: multiple feature channels, feature extraction, and multistage cascade AdaBoost

Table 3 Multilayer image channels. The first layer is HOG+LUV, and the remaining layers are the convolutional layers (i.e., C1 to C5) in VGG16

Layer	L1	L2	L3	L4	L5	L6
Name	HOG	VGG16				
	LUV	C1	C2	C3	C4	C5
Size	128 × 64	64 × 32	32 × 16	16 × 8	8 × 4	4 × 2
Num	10	64	128	256	512	512

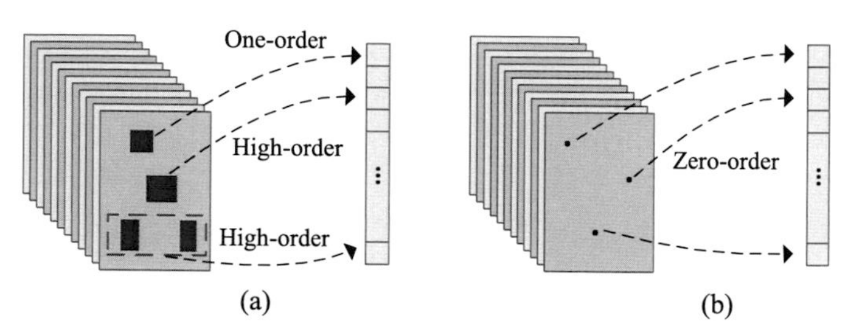

Fig. 21 Feature extraction in L1 and L2–L6. (**a**) is NNNF for L1, (**b**) is the single pixel for L2–L6

example, a five-layer feature channels can be generated by HOG+LUV and C2–C5 of VGG16. C1 of VGG16 is not used.

Feature extraction methods used in multilayer feature channels are different. For L1 (i.e., HOG+LUV), many successful methods can been proposed (e.g., ACF [6, 16, 43, 72, 73]). ACF [16] and NNNF [6] are selected for feature extraction in L1. ACF [16] has a very fast detection speed, while NNNF [6, 7] has a very good detection speed performance and relatively fast detection speed. For L2–L6 (CNN channels), the number of channels is relatively large. If the relative complexity features are used, the computation cost will be large. Thus, only the single pixel value is used as the candidate feature in L2–L6 for the computation efficiency. Figure 21 shows the feature extraction in L1 and L2–L6.

Based on multilayer image channels and candidate features extracted in each layer, multistage cascade AdaBoost is used to learn the strong classifier. Rows 2–4 in Fig. 15 shows the explanations of multistage cascade AdaBoost. The weak classifiers in each stage are learned from the candidate features extracted from corresponding feature channels. The strong classifier (i.e., $H(\mathbf{x})$) of multistage cascade AdaBoost can be written as follows:

$$H(\mathbf{x}) = \sum_{j=1}^{k_1} \alpha_1^j h_1^j(\mathbf{x}) + \ldots + \sum_{j=1}^{k_i} \alpha_i^j h_i^j(\mathbf{x}) + \ldots + \sum_{j=1}^{k_N} \alpha_N^j h_N^j(\mathbf{x}) = \sum_{i=1}^{N} \sum_{j=1}^{k_i} \alpha_i^j h_i^j(\mathbf{x}), \quad (23)$$

where $\mathbf{x}$ means the detection windows, $h_i^j(\mathbf{x})$ is the j-th weak classifier in stage i, α_i^j is th weight of $h_i^j(\mathbf{x})$, and $k_1, k_2, \ldots, k_N$ are the number of weak classifiers in each stage. How to set $k_1, k_2, \ldots, k_N$ is an opening problem. Generally speaking, one empirical settings are used as follows:

$$k_1 = N_{All}/2,$$
$$k_2 = k_3 = \ldots = k_N = N_{All}/(2 \times (N - 1)), \quad (24)$$

where N_{All} is the total number of weak classifiers in the strong classifier, which is usually set as 2048 or 4096.

Figure 22 shows the test process of MCF. In the test stage, the channels of HOG+LUV are firstly computed. Detection widows are generated by scanning the whole input image. These detection windows are firstly classified by the classifier of S1. For detection windows accepted by S1, the channels of L2 are then computed, and these windows are classified by S2. The above process is repeated from L1 to LN. Finally, the detection windows accepted by all the stages of strong classifier (i.e., $H(\mathbf{x})$) are merged by NMS as the final pedestrian detection result. Generally, detection windows around pedestrians highly overlap; they need much computation cost. Thus, the highly overlap detection windows after the first stage are eliminated by NMS with the threshold 0.8. It can further accelerate detection speed with little performance loss.

Table 4 shows miss rates (MR) of MCF-based HOG+LUV and different layers in VGG16 on Caltech test set. They are trained on Caltech10x training set with 4096 level-2 decision forests. $\surd$ means that the corresponding layer is used. For example, MCF-2 is constructed by HOG+LUV and C5 of VGG16. It can be

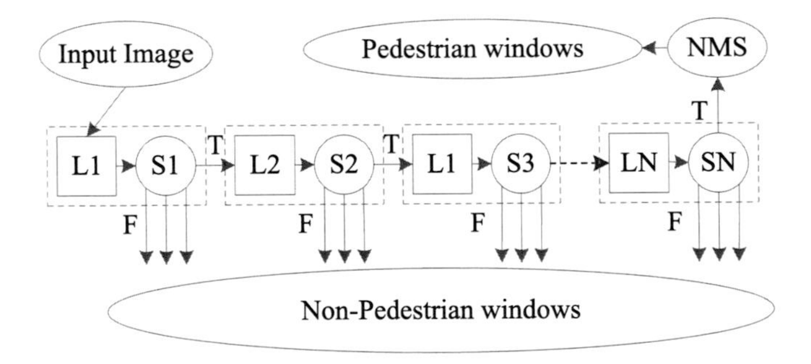

Fig. 22 The test process of MCF

Table 4 Miss rates (MR) of MCF based on HOG+LUV and the different layers in CNN on Caltech test set. $\surd$ means that the corresponding layer is used. HOG+LUV is always used for the first layer. The layers in VGG16 are used for the remaining layers

	HOG	VGG16						
Name	LUV	C1	C2	C3	C4	C5	MR (%)	Δ MR (%)
MCF-2	√					√	18.52	N/A
MCF-3	√				√	√	17.14	1.38
MCF-4	√			√	√	√	15.40	3.12
MCF-5	√		√	√	√	√	14.78	3.74
MCF-6	√	√	√	√	√	√	14.31	4.21

seen that with more convolutional layers, MCF has the lower miss rate. MCF-6 has the best detection performance, which outperforms MCF-2 and MCF-5 by 4.21% and 1.47%, respectively. It demonstrates that the middle layers in CNN can enrich the feature abstraction. Based on each layer in CNN, MCF can extract more discriminative features, which can be used for pedestrian detection.

Table 5 compares miss rates and detection times of MCF-2, MCF-6, and MCF-6-f. MCF-6-f eliminates the highly overlapped detection windows by NMS with the threshold of 0.8. The miss rate of MCF-6-f is 4.21% lower than that of MCF-2, while the detection speed of MCF-6 is 1.43 times faster than that of MCF-2. By eliminating the highly overlapped detection windows after stage 1, MCF-6-f can further accelerate detection speed with little detection performance loss (i.e., 4.21%). The miss rate of MCF-6-f is 3.63% low; the detection speed of MCF-6-f is 4.07 times faster than that of MCF-2. Because MCF can reject many detection windows by first few stages, it can accelerate detection speed.

To compare with state-of-the-art performance, MCF is trained on Caltech10x training set by 4096 level-4 decision forests. Figure 23 compares MCF with some state-of-the-art methods (i.e., LatSvm [19], ACF [16], LDCF [43], Checkerboards [73], CCF+CF [66], DeepParts [58], and CompACT-Deep [4]) on Caltech test set. It can be seen that MCF achieves state-of-the-art performance, which outperforms DeepParts and CompACT-Deep by 1.49% and 1.35%, respectively.

Moreover, MCF is further trained on KITTI training set by 4096 level-4 decision forests. Table 6 compares MCF with some state-of-the-art methods (i.e., ACF [16], SpatialPooling+ [46], Checkerboards [73], DeepParts [58], and CompACT-Deep [4]) on KITTI test set. MCF also outperforms DeepParts and CompACT-Deep. On the moderate test, MCF outperforms CompACT-Deep by 0.71%. On the hard test set, MCF outperforms CompACT-Deep by 1.57%.

Table 5 Miss rate (MR) and detection time of MCF-2, MCF-6, and MCF-6-f. MCF-2 is based on HOG+LUV and C5 in CNN. MCF-6 is based on HOG+LUV and C1–C5 in CNN. MCF-6-f is the fast version of MCF-6 by eliminating overlapped detection windows

	HOG+LUV and VGG16		
	MCF-2	MCF-6	MCF-6-f
MR (%)	18.52	14.31	14.89
Time (s)	7.69	5.37	1.89

Table 6 Average precision (AP) of some methods on KITTI

Method	Easy	Moderate	Hard
ACF [16]	44.49%	39.81%	37.21%
SpatialPooling+ [46]	65.26%	54.49%	48.60%
Checkerboards [73]	67.75%	56.75%	51.12%
DeepParts [58]	70.49%	58.67%	52.78%
CompACT-Deep [4]	70.69%	58.74%	52.71%
MCF	70.87%	59.45%	54.28%

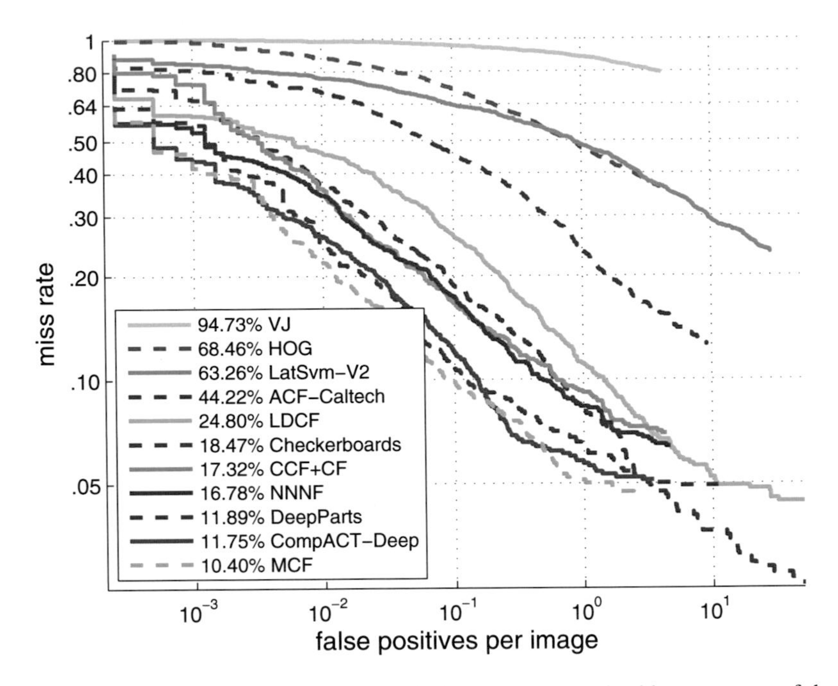

Fig. 23 ROC on Caltech test set (reasonable). MCF is compared with some state-of-the-art methods

4 Challenges of Object Detection

Though object detection has achieved great progress in the past decades, it still has many challenges when pushing the progress of object detection. In the following part, three common and typical challenges of object detection will be discussed, and some solutions are also introduced.

4.1 Scale Variation Problem

As the distance from objects to camera can be various, objects of various scales usually appear on the image. Thus, scale variation is an inevitable problem for object detection. The solutions to scale variation can be divided into two main classes: (1) image pyramid-based methods and (2) feature pyramid-based methods. Generally, image pyramid-based methods firstly resize the original image into multiple different scales and then use the same detector to detect the rescaled images, respectively. Feature pyramid-based methods firstly generate multiple feature maps of different resolution based on the input image and then use different feature maps to detect objects of different scales.

At first, deep object detection adopts image pyramid to detect objects of various scales. RCNN [23], SPPnet [25], Fast RCNN [24], and Faster RCNN [52] all adopt image pyramid for object detection. In the training stage, the CNN detector is trained based on the images of a given scale. In the test stage, image pyramids are used for multi-scale object detection. On the one hand, it usually causes the inconsistency between the training and test inference. On the other hand, each image of image pyramids is put into the CNN network, respectively. Thus, it is also very time-consuming.

In fact, the feature maps of different resolutions in CNN can be seen as the feature pyramid. If the feature maps of different resolutions are used to detect objects of different scales, it can avoid resizing the input image and accelerating detection speed. Thus, feature pyramid-based methods become popular. Researchers have done many attempts on feature pyramid-based methods.

Li et al. [34] proposed scale-aware Fast RCNN (SAF RCNN) for pedestrian detection. The base network is split into two sub-networks for large-scale pedestrian detection and small-scale pedestrian detection, respectively. Given a detection window, the final detection score is the weight sum of two sub-networks. If the detection window is relatively large, the large-scale network has the relatively large weight. If the detection window is relatively small, the small-scale network has the relatively large weight.

Yang et al. [68] proposed scale-dependent pooling (SDP) for multi-scale object detection to handle the scale variation problem. It is based on Fast RCNN architecture. The proposals are extracted by selective search method [60]. According to the heights of proposals, SDP pools the features of proposals from different convolutional layers according to the height of proposals. If the height of object proposal belongs to [0, 64], SDP pools the feature maps from the third convolutional blocks. If the height of object proposal belongs to [64, 128], SDP pools the feature maps from the fourth convolutional blocks. If the height of object proposal belongs to $[128, +inf]$, SDP pools the feature maps from the fifth convolutional blocks. Due to the feature maps of ROI pooling layer are pooled from different convolutional layers, three different subnets for classifying and locating proposals are trained, respectively.

Generally, Faster RCNN needs to extract proposals by sliding RPN on a fixed convolutional layer (e.g., conv5_3 of VGG16). Because the respective field of a convolutional layer is relatively fixed, it cannot match the sizes of all objects very well. The respective field of former convolutional layer is relatively small, which matches the small-scale objects better, while the respective field of latter convolutional layer is relatively large, which matches the large-scale objects better. To solve this problem, Cai et al. [5] proposed multi-scale deep convolutional neural network (MS-CNN) to generate object proposals of different scales. Figure 24 shows the architecture of MS-CNN. It outputs proposals from multiple convolutional layers of different resolution. The anchor in each convolutional layer is labelled as the positive if it has an IoU overlap over 0.5 with a ground-truth bounding box. The anchor in each convolutional layer is labelled as the negative if it has the IoU overlap below 0.2 with all the ground-truth bounding boxes. Because of the imbalance

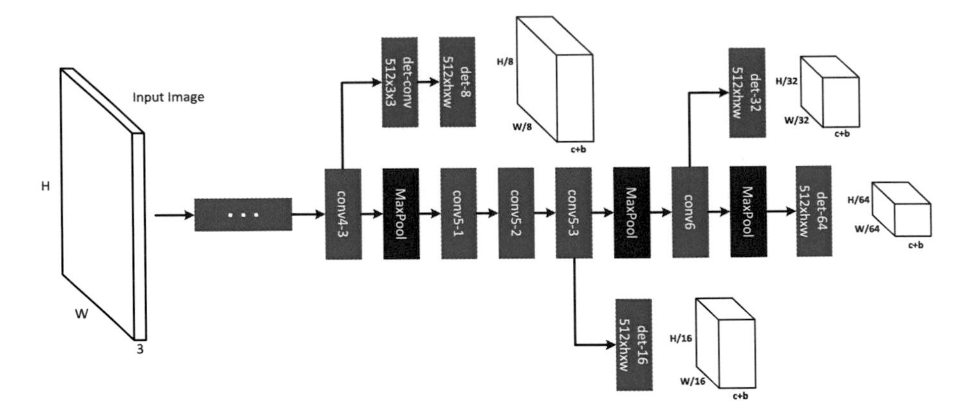

Fig. 24 The architecture of MS-CNN. (Reprinted from Ref. [5], with permission of Springer)

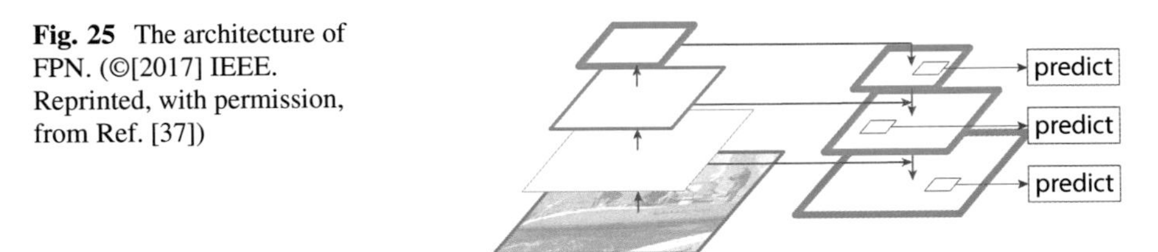

Fig. 25 The architecture of FPN. (©[2017] IEEE. Reprinted, with permission, from Ref. [37])

between the negatives and positives, three different sampling strategies are explored: random, bootstrapping, and mixture. Random sampling strategy randomly selects some negatives from all the negatives. Bootstrapping sampling strategy selects some hardest negatives according to their objectness scores. Mixture sampling strategy selects the half negatives by random sampling strategy and the half negatives by bootstrapping strategy. It is found that the bootstrapping sampling strategy and mixture sampling strategy have similar detection performance, which outperforms random sampling strategy. The ratio of positives and negatives is 1:R.

Though the success of MS-CNN, the output layers of MS-CNN have different semantic levels. To solve the semantic inconstancy of different convolutional layers, Lin et al. [37] proposed feature pyramid network (FPN) for object detection. FPN incorporates top-down structure to improve the semantic levels of first few convolutional layers. Figure 25 shows the architecture of FPN. Specifically, top-down structure combines the low-level semantic convolutional layer with the upsampled high-level semantic convolutional layer by the element-wise addition. Based on FPN, the semantics of output layers for proposal generation are high-level and consistent. This final feature maps used for predicting proposals are called $\{P2, P3, P4, P5\}$, corresponding to the convolutional layers of $\{C2, C3, C4, C5\}$. The anchors have areas of $\{32 \times 32,\ 64 \times 64,\ 128 \times 128,\ 256 \times 256,\ 512 \times 512\}$ pixels on $\{P2, P3, P4, P5\}$, respectively. The aspect ratios of anchors are $\{1:2,\ 1:1,\ 2:1\}$. Thus, there are 15 anchors over the pyramid. The anchor is labelled as the positive if it has the highest IoU overlap with a ground-truth box or if it has the IoU overlap over 0.7 with any ground-truth boxes. The anchor is labelled as the negative if it has the IoU overlap below 0.3 for all the ground-truth boxes. With consistency feature maps for object detection, it can improve detection performance, especially the detection performance of small-scale object detection. The similar idea is also adopted by [32].

Because small-scale objects are usually low-resolution and noisy, small-scale object detection is more challenging compared to large-scale object detection. Though multi-scale methods treat objects of different scales as different classes, the improvement is relatively limited. Thus, improving small-scale object detection is a key for multi-scale object detection. To solve this problem, Li et al. [35] proposed the perceptual generative adversarial network (Perceptual GAN). Based on the difference between feature representations of small-scale objects and feature representations of large-scale objects, perceptual GAN aims to compensate feature representations of small-scale objects.

Generally, large-scale objects have more abundant information compared to small-scale object. To improve small-scale pedestrian detection, Pang et al. [47] proposed JCS-Net. The main idea of JCS-Net is to use large-scale pedestrian to help small-scale pedestrian detection. Figure 26 shows the pipeline of JCS-Net. The training process of JCS-Net for small-scale pedestrian can be summarized as follows: (1) It firstly fine-tunes the network for large-scale pedestrian detection based on large-scale pedestrians (i.e., the top row of Fig. 26). (2) The super-resolution sub-network for small-scale pedestrian detection (i.e., the left of bottom row of Fig. 26) is pre-trained by the large-scale pedestrians and corresponding small-scale pedestrians. (3) The classification sub-network for small-scale pedestrian detection (i.e., the right of bottom row of Fig. 26) is initialized by the large-scale network. (4) The super-resolution sub-network and classification sub-network for small-scale pedestrian detection are jointly trained based on small-scale pedestrians and corresponding negatives.

The loss of JCS-Net is the joint of two sub-networks. The loss of super-resolution sub-network can be calculated by mean squared error as:

$$L_{similarity} = \frac{1}{n}\sum_{i=1}^{n} ||\mathbf{y}_i - F(\mathbf{x}_i)||^2, \tag{25}$$

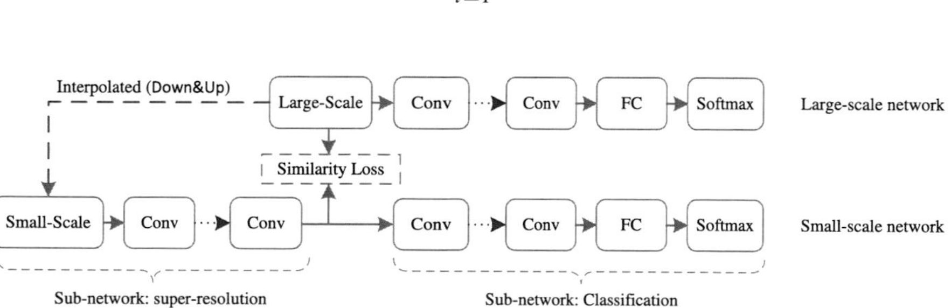

Fig. 26 The architecture of JCS-Net

where $\mathbf{y}_i$ is large-scale pedestrian, $\mathbf{x}_i$ is small-scale pedestrian, $F(\mathbf{x}_i)$ is the reconstructed pedestrian by super-resolution sub-network, and n is the number of training samples. The loss of classification can be written as:

$$L_{cls} = \frac{1}{n}\sum_{i=1}^{n} - \log p_c(\mathbf{x}_i), \tag{26}$$

where c is the ground-truth label of $\mathbf{x}_i$ and $p_c(\mathbf{x}_i)$ is the probability that $\mathbf{x}_i$ belongs to class c. Based on $L_{similarity}$ and L_{cls}, the joint loss of JCS-Net (i.e., $LJCS$) can be expressed as:

$$L_{JCS} = L_{cls} + \lambda L_{similarity}, \tag{27}$$

where λ is used to balance two terms (i.e., L_{cls} and $L_{similarity}$) which is set to be 0.1 by cross-validation.

To detect multi-scale pedestrians, multi-scale MCF can be trained based on JCS-Net or original HOG+LUV and VGG16. Generally, MCF-V is trained based on original HOG+LUV and VGG16, which is used for large-scale pedestrian detection. MCF-J is trained based on JCS-Net, which is used for small-scale pedestrian detection. Figure 27 gives an illustration of multi-scale MCF. Pedestrians of different scales are divided into several different subsets (i.e., subset 1, subset 2, ..., subset N) according to the height of pedestrians. The different subsets can contain some overlapped images. In Fig. 27, two detectors on subset 1 and subset 2 are trained based on MCF-J. The other detectors are trained based on MCF-V. In the test stage, the detection results of each detector are added together before NMS.

The original pedestrians on Caltech training set are split into three different subsets, which are called "train-all," "train-small," and "train-large," respectively.

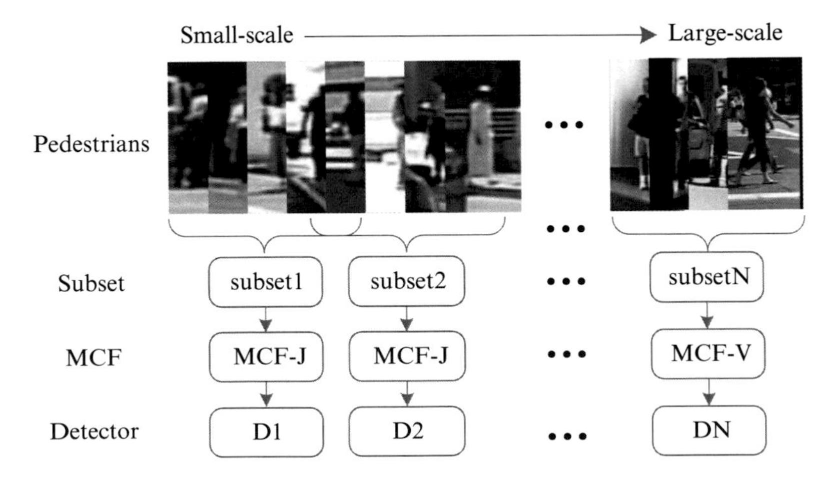

Fig. 27 The illustration of multi-scale MCF

"Train-all" subset contains all the pedestrians, "train-small" subset contains the pedestrians under 100 pixels tall and the interpolated pedestrians over 100 pixels tall, and "train-large" subset contains the pedestrians over 80 pixels tall. The two subsets of Caltech test set (i.e., reasonable and small) are used for evaluation. The reasonable test set means that pedestrians are over 50 pixels tall under no or partial occlusion, and the small test set means that pedestrians are under 100 pixels tall and over 50 pixels tall. Namely, the small test set belongs to the reasonable test set.

Table 7 shows the miss rates of MCF-V and MCF-J trained on "train-small" training set. MCF-J outperforms MCF-V on the reasonable test set and the small test set, especially on the small test set. Two ablation experiments are also conducted: (1) The first one is that setting λ of (27) is 0, which aims to show the influence of depth of JCS-Net. It is called MCF-C. (2) The second one is that the super-resolution sub-network and classification sub-network are not jointly trained, which aims to show the importance of the joint multitask training. It is called MCF-S. Both MCF-C and MCF-S are superior to MCF-V and inferior to MCF-J. It means that though deeper depth and simple super-resolution can improve detection performance, the joint multitask training of JCS-Net is important, which can further improve detection performance.

Based on three training sets (i.e., "train-all," "train-small," and "train-large"), MS-V and MS-J are trained. They both contain three different detectors. The difference is that on "train-small" training set, MS-V uses MCF-V, and MS-J uses MCF-J. Table 8 shows the miss rates of MS-V and MS-J on Caltech test set. The miss rates of MS-J have 0.86% and 0.91% lower than that of MS-V on reasonable test set and small test set. It demonstrates the effectiveness of MS-J.

Finally, MS-J is compared to some state-of-the-art methods (i.e., Roerei [2], ACF [16], LDCF [43], TA-CNN [59], Checkerboards [73], DeepParts [58], CompACT-

Table 7 Miss rates (MR) of MCF-V and MCF-J are shown on Caltech test set. MCF-V is learned based on HOG+LUV and the fine-tuned VGG16. MCF-J is learned based on HOG+LUV and the proposed JCS-Net

Method	Training set	Reasonable	Small
MCF-V	"train-small"	13.20%	14.28%
MCF-J	"train-small"	11.07%	11.72%
ΔMR	–	2.13%	2.56%
Ablation experiments			
MCF-C	"train-small"	12.23%	13.02%
MCF-S	"train-small"	12.65%	13.50%

Table 8 Miss rates (MR) of MS-V and MS-J are shown on Caltech test set. MS-V means multi-scale MCF based on fine-tuned VGG16. MS-J means multi-scale MCF based on JCS-Net

Method	Detectors	Training set	Reasonable	Small
	MCF-V	"train-small"		
MS-V	MCF-V	"train-large"	9.67%	10.48%
	MCF-V	"train-all"		
	MCF-J	"train-small"		
MS-J	MCF-V	"train-large"	8.81%	9.57%
	MCF-V	"train-all"		
ΔMR	–	–	0.86%	0.91%

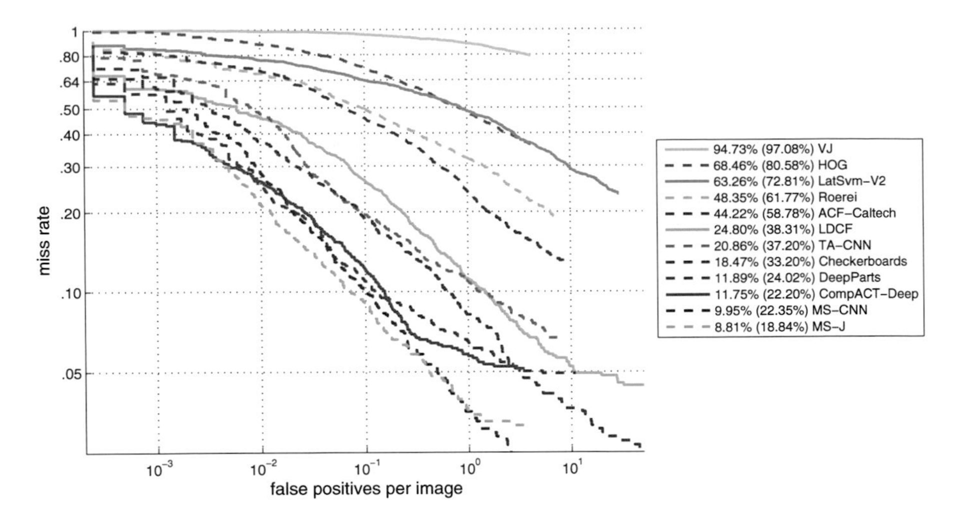

Fig. 28 ROC on Caltech test set. MS-J is compared to some state-of-the-art methods

Deep [4], and MS-CNN [5]) on Caltech test set. Figure 28 shows the ROC. MS-J achieves state-of-the-art performance, which outperforms MS-CNN by 1.14%.

4.2 *Occlusion Problem*

Object occlusion is very common. For example, in [15], Dollar et al. found that most pedestrians (about 70%) in street scenes are occluded in at least one frame. Thus, detecting occluded object is very necessary and important for computer vision application. In the past decade years, researchers have done many attempts to solve occlusion problem.

Wang et al. [62] found that if some parts of pedestrian are occluded, the block features of corresponding region uniformly respond to the block scores of linear classifier. Based on this phenomenon, they proposed to use the score of each block to judge whether the corresponding region is occluded or not. Based on the scores of each block, the occlusion likelihood images are segmented by the mean shift approach. If occlusion occurs, the part detector is applied on the unoccluded regions to output the final detection result.

To maximize detection performance on the occluded pedestrians, Mathias et al. [40] proposed to learn a set of occlusion-specific pedestrian detectors. Each pedestrian detector serves for the occlusion of certain type. In [40], occlusion can be divided into three different types: occlusions from bottom, occlusions from right, and occlusions from left. For each type, the degree of occlusion ranges from 0% to 50%. Eight left/right occlusion detectors and 16 bottom-up occlusion detectors are trained, respectively. One nave method to obtain these classifiers is to train the classifiers of all the occlusion levels, respectively. However, it is very time-consuming. To reduce the training time, Franken-classifiers are proposed. It starts to train the full-body biased classifier and remove weak classifiers to generate the first occlusion classifier. The additional weak classifiers of first occlusion classifier are further learned without bias. Similar to the first occlusion classifier, the second occlusion classifier are learned based on the first occlusion classifier. Based on Franken-classifiers, it only needs one tenth computation cost for training the set of occlusion-specific pedestrian detectors.

Inspired by the set of occlusion-specific pedestrian detectors, Tian et al. [58] extended it by constructing an extensive deep part pool and automatically choose important parts for occlusion handling by linear SVM. The extensive part pool contains various body parts. Pedestrians can be seen as a rigid object, which are divided into the $2m \times m$ spatial grids. The part pools consist of all the rectangles inside the spatial grids, where the height and width of the rectangle are at least 2. If $m = 3$, the part pool has 45 part models. To alleviate the test computation cost of 45 part models, 6 part models with highest SVM scores are selected, which also yield approximate performance with faster detection speed.

Wang et al. [64] thought that data-driven strategy is very important for solving occlusion problem. If the training data has enough image data of all different occlusion situations, the training detector can have a better detection performance for occluded objects. However, the dataset cannot cover all the cases of occlusions generally, and the occlusions are relatively rare. Because the occlusions in the dataset follow a long-tail distribution, it is impossible to collect the dataset to cover all the occlusions. To solve this problem, Wang et al. proposed A-Fast-RCNN, which uses adversary network to generate hard examples by blocking some feature maps spatially. After ROI pooling layer, it adds an extra branch to generate the occlusion mask. The branch consists of two fully connected layers and mask prediction layer. The feature maps for the final classification and regression are the combination of mask and the feature map of ROI pooling layer. If the cell of mask is equal to 1, the corresponding responses of feature maps are set as 0 (i.e., dropout). In the training stage, it adopts the stage-wise steps: (1) The occlusion mask is fixed; Fast RCNN network is firstly trained. (2) Based on Fast RCNN, the adversary network generates the occlusion mask which makes the loss of Fast RCNN the largest. Through the two-stage training, the final Fast RCNN network can be robust to the occlusion. In the test stage, the mask branch is removed, and the rest process is the same as the original Fast RCNN.

4.3 Deformation Problem

Object deformation can be caused by non-grid deformation, intra-class shape variability, and so on. For example, people can jump or squat. Thus, a good object detection method should be robust to the deformation of object. Before CNN, researchers have done many attempts to handle deformation. For example,

DPM [19] uses the mixtures of multi-scale deformable part models (i.e., one low-resolution root model and six high-resolution part models) to handle deformation. HSC [51] further incorporates histograms of sparse codes into deformable part model. To accelerate detection speed of DPM, CDPM [18] and FDPM [65] are further proposed. Park et al. [48] proposed to detect large-scale pedestrians by deformable part model and detect small-scale pedestrians by the rigid template. Regionlets [63] presents the region by a set of small sub-regions with different sizes and aspect ratios.

Though CNN-based methods are robust to object deformation in some degree, it is still not good enough. To further improve the robustness to object detection, researchers also incorporate some specific design into the CNN-based methods. Ouyang et al. [45] proposed the deformation constrained pooling layer (def-pooling) to model the deformation of object parts. Traditional pooling (e.g., max-pooling and average pooling) can be replaced by def-pooling to better represent the deformation properties of objects. Recently, Dai et al. [12] proposed two deformable modules (i.e., deformable convolution and deformable ROI pooling) to enhance the representation ability of geometric transformation. They add 2D offset to the regular grid sampling locations in the standard convolution. The offsets are learned in the training stage. Jeon and Kim [30] also proposed active convolution, where the shape of convolution is learned in the training stage. To improve invariance to large deformation and transformation, Jaderberg et al. [29] proposed the spatial transformer network. The transformation of scaling, rotation, and non-rigid deformations is performed on the feature map.

References

1. Bell, S., Zitnick, C. L., Bala, K., and Girshick, R.: Inside-Outside Net: Detecting objects in context with skip pooling and recurrent neural networks. in Proc. IEEE Conf. Computer Vision and Pattern Recognition (2016)
2. Benenson, R., Mathias, M., Tuytelaars, T., and Gool, L. V.: Seeking the strongest rigid detector. in Proc. IEEE Conf. Computer Vision and Pattern Recognition (2013)
3. Benenson, R., Omran, M., Hosang, J. and Schiele, B.: Ten years of pedestrian detection, what have we learned? in Proc. Eur. Conf. Comput. Vis. (2014)
4. Cai, Z., Saberian, M., and Vasconcelos, N.: Learning complexity-aware cascades for deep pedestrian detection. in Proc. IEEE Int. Conf. Comput. Vis. (2015)
5. Cai, Z. Fan, Q., Feris, R. S., and Vasconcelos, N.: A unified multi-scale deep convolutional neural network for fast object detection. in Proc. Eur. Conf. Comput. Vis. (2016)
6. Cao, J., Pang, Y., and Li, X.: Pedestrian detection inspired by appearance constancy and shape symmetry. in Proc. IEEE Conf. Computer Vision and Pattern Recognition (2016)
7. Cao, J., Pang, Y., and Li, X.: Pedestrian detection inspired by appearance constancy and shape symmetry. IEEE Trans. Image Processing 25(12), 5538–5551 (2016)
8. Cao, J., Pang, Y., and Li, X.: Learning multilayer features for pedestrian detection. IEEE Trans. Image Processing 26(7), 3310–3320 (2017)
9. Cheng, M. M., Zhang, Z., Lin, W.Y., and Torr, P.: BING: Binarized normed gradients for objectness estimation at 300fps. in Proc. IEEE Conf. Computer Vision and Pattern Recognition (2014)

10. Dalal, N. and Triggs, B.: Histograms of oriented gradients for human detection. Int. J. Comput. Vis. 57(2), 137–154 (2004)
11. Dai, J., Li, Y., He, K., and Sun, J.: R-FCN: Object detection via region-based fully convolutional networks. in Proc. Advances in Neural Information Processing Systems (2016)
12. Dai, J., Qi, H., Xiong, Y., Li, Y., Zhang, G., Hu, H., and Wei, Y.: Deformable convolutional networks. in Proc. IEEE Conf. Computer Vision and Pattern Recognition (2017)
13. Deng, J., Dong, W., Socher, R., Li, L.-J., Li, K., and Fei-Fei, L.: ImageNet: A large-scale hierarchical image database. in Proc. IEEE Conf. Computer Vision and Pattern Recognition (2009)
14. Dollár, P., Tu, Z., Perona, P., and Belongie, S.: Integral channel features. in Proc. Brit. Mach. Vis. Conf. (2009)
15. Dollár, P., Wojek, C., Schiele, B., and Perona, P.: Pedestrian detection: An evaluation of the state of the art. IEEE Trans. Pattern Analysis and Machine Intelligence 34(4), 743–761 (2012)
16. Dollár, P., Appel, R., Belongie, S., and Perona, P.: Fastest feature pyramids for object detection. IEEE Trans. Pattern Analysis and Machine Intelligence 36(8), 1532–1545 (2014)
17. Everingham, M., Van Gool, L.,Williams, C. K. I.,Winn, J., and Zisserman, A.: The PASCAL visual object classes (VOC) challenge. Int. J. Comput. Vis. 88(2), 303–338 (2010)
18. Felzenszwalb, P. F., Girshick, R., and McAllester, D.: Cascade object detection with deformable part models. in Proc. IEEE Conf. Computer Vision and Pattern Recognition (2010)
19. Felzenszwalb, P., Girshick, R., McAllester, D., and Ramanan, D.: Object detection with discriminatively trained part based models. IEEE Trans. Pattern Analysis and Machine Intelligence 32(9), 1627–1645 (2012)
20. Fu, C.-Y., Liu, W., Ranga, A., Tyagi, A., and Berg, A. C.: Dssd: Deconvolutional single shot detector. CoRR abs/1701.06659 (2017)
21. Geiger, A., Lenz, P., and Urtasun, R.: Are we ready for autonomous driving? The KITTI vision benchmark suite. in Proc. IEEE Conf. Computer Vision and Pattern Recognition (2012)
22. Gidaris, S. and Komodakis, N.: LocNet: Improving localization accuracy for object detection. in Proc. IEEE Conf. Computer Vision and Pattern Recognition (2016)
23. Girshick, R., Donahue, J., Darrell, T., and Malik, J.: Rich feature hierarchies for accurate object detection and semantic segmentation. in Proc. IEEE Conf. Computer Vision and Pattern Recognition (2014)
24. Girshick, R.: Fast RCNN. in Proc. Int. Conf. Comput. Vis. (2015)
25. He, K., Zhang, X., Ren, S., and Sun, J.: Spatial pyramid pooling in deep convolutional networks for visual recognition. IEEE Trans. Pattern Analysis and Machine Intelligence 37(9), 1904–1916 (2015)
26. He, K., Zhang, X., Ren, S., and Sun, J.: Deep residual learning for image recognition. in Proc. IEEE Conf. Computer Vision and Pattern Recognition (2016)
27. He, K., Gkioxari, G., Dollár, P., and Girshick, R.: Mask R-CNN. in Proc. Int. Conf. Comput. Vis. (2017)
28. Hosang, J., Omran, M., Benenson, R., and Schiele, B.: Taking a deeper look at pedestrians. in Proc. IEEE Conf. Computer Vision and Pattern Recognition (2015)
29. Jaderberg, M., Simonyan, K., Zisserman, A., and Kavukcuoglu, K.: Spatial transformer networks. in Proc. Advances in Neural Information Processing Systems (2015)
30. Jeon, Y. and Kim, J.: Active convolution: Learning the shape of convolution for image classification. in Proc. IEEE Conf. Computer Vision and Pattern Recognition (2017)
31. Kong, T., Yao, A., Chen, Y., and Sun, F.: HyperNet: Towards accurate region proposal generation and joint object detection. in Proc. IEEE Conf. Computer Vision and Pattern Recognition (2016)
32. Kong, T., Sun, F., Yao, A., Liu, H., Lu, M., and Chen, Y.: Ron: Reverse connection with objectness prior networks for object detection. in Proc. IEEE Conf. Computer Vision and Pattern Recognition (2017)
33. Krizhevsky, A., Sutskever, I., and Hinton, G. E.: ImageNet classification with deep convolutional neural networks. in Proc. Advances in Neural Information Processing Systems (2012)

34. Li, J., Liang, X., Shen, S., Xu, T., and Yan, S.: Scale-aware Fast R-CNN for pedestrian detection. CoRR abs/1510.08160 2015
35. Li, J., Liang, X., Wei, Y., Xu, T., Feng, J., and Yan, S.: Perceptual generative adversarial networks for small object detection. in Proc. IEEE Conf. Computer Vision and Pattern Recognition (2017)
36. Lin, T.-Y., Goyal, P., Girshick, R., He, K., and Dollár, P.: Focal loss for dense object detection. in Proc. Int. Conf. Comput. Vis. (2017)
37. Lin, T.-Y., Dollár, P., Girshick, R., He, K., Hariharan, B., and Belongie, S.: Feature pyramid networks for object detection. in Proc. IEEE Conf. Computer Vision and Pattern Recognition (2017)
38. Liu, W., Anguelov, D., Erhan, D., Szegedy, C., Reed, S., Fu, C.-Y., and Berg, A. C.: SSD: Single shot multibox detector. in Proc. Eur. Conf. Comput. Vis. (2016)
39. Lowe, D.G.: Distinctive image features from scale-invariant keypoints. Int. J. Comput. Vis. 60(2), 91–110 (2004)
40. Mathias, M., Benenson, R., Timofte, R., and Van Gool, L.: Handling occlusions with franken-classifiers. in Proc. Int. Conf. Comput. Vis. (2013)
41. Mao, J., Xiao, T., Jiang, Y., and Cao, Z.: What can help pedestrian detection? in Proc. IEEE Conf. Computer Vision and Pattern Recognition (2017)
42. Najibi, M., Rastegari, M., and Davis, L. S.: G-CNN: an iterative grid based object detector. in Proc. IEEE Conf. Computer Vision and Pattern Recognition (2016)
43. Nam, N., Dollár, P., and Han, J.: Local decorrelation for improved detection. in Proc. Advances in Neural Information Processing Systems (2014)
44. Ojala, T., Pietikainen, M., and Maenpaa, T.: Multiresolution gray-scale and rotation invariant texture classification with local binary patterns. IEEE Transactions on Pattern Analysis and Machine Intelligence 24(7), 971–987 (2002)
45. Ouyang, W., Wang, X., Zeng, X., Qiu, S. Luo, P., Tian, Y., Li, H., Yang, S., Wang, Z., Loy, C.-C., and Tang, X.: DeepID-Net: Deformable deep convolutional neural networks for object detection. in Proc. IEEE Conf. Computer Vision and Pattern Recognition (2015)
46. Paisitkriangkrai, S., Shen, C., and van den Hengel, A.: Pedestrian detection with spatially pooled features and structured ensemble learning. IEEE Trans. Pattern Analysis and Machine Intelligence. 38(6), 1243–1257 (2016)
47. Pang, Y., Cao, J., and Shao, L.: Small-scale pedestrian detection by joint classification and super-resolution into a unified network. Tech. report (2017)
48. Park, D., Ramanan. D., and Fowlkes, C.: Multiresolution models for object detection. in Proc. Eur. Conf. Comput. Vis. (2010)
49. Redmon, J., Divvala, S., Girshick, R., and Farhadi, A.: You only look once: unified, real-time object detection. in Proc. IEEE Conf. Computer Vision and Pattern Recognition (2016)
50. Ren, J., Chen, X., Liu, J., Sun, W., Pang, J., Yan, Q., Tai, Y., and Xu, L:. Accurate single stage detector using recurrent rolling convolution. in Proc. IEEE Conf. Computer Vision and Pattern Recognition (2017)
51. Ren, X., and Ramanan, D.: Histograms of sparse codes for object detection. in Proc. IEEE Conf. Computer Vision and Pattern Recognition (2016)
52. Ren, S., He, K., Girshick, R., and Sun, J.: Faster R-CNN: Towards real-time object detection with region proposal networks. in Proc. Advances in Neural Information Processing Systems (2015)
53. Sermanet, P., Kavukcuoglu, K., Chintala, S., and LeCun, Y.: Pedestrian detection with unsupervised multi-stage feature learning. in Proc. IEEE Conf. Computer Vision and Pattern Recognition (2013)
54. Sermanet, P., Eigen, D., Zhang, X., Mathieu, M., Fergus, R., and LeCun, Y.: Overfeat: Integrated recognition, localization and detection using convolutional networks. In ICLR, (2014)
55. Shrivastava, A., Gupta, A., and Girshick, R.: Training region-based object detectors with online hard example mining. in Proc. IEEE Conf. Computer Vision and Pattern Recognition (2016)

56. Simonyan, K., and Zisserman, A.: Very deep convolutional networks for large-scale image recognition. CoRR abs/1409.1556 (2014)
57. Szegedy, C., Liu, W., Jia, Y., Sermanet, P., Reed, S., Anguelov, D., Erhan, D., Vanhoucke, V., and Rabinovich, A.: Going deeper with convolutions. in Proc. IEEE Conf. Computer Vision and Pattern Recognition (2015)
58. Tian, Y., Luo, P., Wang, X., Tang, X.: Deep learning strong parts for pedestrian detection. in Proc. Int. Conf. Comput. Vis. (2015)
59. Tian, Y., Luo, P., Wang, X., and Tang, X.: Pedestrian detection aided by deep learning semantic tasks. in Proc. IEEE Conf. Computer Vision and Pattern Recognition (2015)
60. Uijlings, J. R. R., van de Sande, K. E. A., Gevers, T., and Smeulders, A. W. M.: Selective search for object recognition. Int. J. Comput. Vis. (2013)
61. Viola, P. and Jones, M.J.: Robust real-time face detection. Int. J. Comput. Vis. 57(2), 137–154 (2004)
62. Wang, X., Han, T. X., and Yan, S.: An HOG-LBP human detector with partial occlusion handling. in Proc. IEEE Conf. Computer Vision and Pattern Recognition (2008)
63. Wang, X., Yang, M., Zhu, S., and Lin, Y.: Regionlets for generic object detection. in Proc. Int. Conf. Comput. Vis. (2013)
64. Wang, X., Shrivastava, A., and Gupta, A.: A-Fast-RCNN: Hard positive generation via adversary for object detection. in Proc. Int. Conf. Comput. Vis. (2017)
65. Yan, J., Lei, Z., Wen, L., and Li, S. Z.: The fastest deformable part model for object detection. in Proc. IEEE Conf. Computer Vision and Pattern Recognition (2014)
66. Yang, B., Yan, J., Lei, Z., and Li, S. Z.: Convolutional channel features. in Proc. Int. Conf. Comput. Vis. (2015)
67. Yang, B., Yan, J., Lei, Z., Li, and S. Z.: CRAFT objects from images. in Proc. IEEE Conf. Computer Vision and Pattern Recognition (2016)
68. Yang, F., Choi, W., and Lin, Y.: Exploit All the Layers: Fast and accurate CNN object detector with scale dependent pooling and cascaded rejection classifiers. in Proc. IEEE Conf. Computer Vision and Pattern Recognition (2016)
69. Yu, F., Koltun, V., and Funkhouser, T.: Dilated residual networks. in Proc. IEEE Conf. Computer Vision and Pattern Recognition (2017)
70. Zagoruyko, S., Lerer, A., Lin, T.-Y., Pinheiro, P. O., Gross, S., Chintala, S., and Dollár, P.: A multipath network for object detection. in Proc. British Machine Vision Conference (2016)
71. Zhang, L., Lin, L., Liang, X., and He, K.: Is faster R-CNN doing well for pedestrian detection? in Proc. Eur. Conf. Comput. Vis. (2016)
72. Zhang, S., Bauckhage, C., and Cremers, A. B.: Informed haar-like features improve pedestrian detection. in Proc. IEEE Conf. Computer Vision and Pattern Recognition (2014)
73. Zhang, S., Benenson, R., and Schiele, B.: Filtered channel features for pedestrian detection, in Proc. IEEE Conf. Computer Vision and Pattern Recognition (2015)
74. Zhang, S., Benenson, R., Hosang, J. and Schiele, B.: CityPersons: A diverse dataset for pedestrian detection. in Proc. IEEE Conf. Computer Vision and Pattern Recognition (2016)
75. Zitnick, C. L., and Dollár, P.: Edge boxes: locating object proposals from edges. in Proc. Eur. Conf. Comput. Vis. (2014)

Deep Learning in Face Recognition Across Variations in Pose and Illumination

Xiaoyue Jiang, Yaping Hou, Dong Zhang, and Xiaoyi Feng

Abstract Even though face recognition in frontal view and normal lighting conditions works very well, the performance drops sharply in extreme conditions. Recently there is plenty of work dealing with pose and illumination problems, respectively. However both the lighting and pose variations always happen simultaneously in general conditions, and consequently we propose an end-to-end face recognition algorithm to deal with two variations at the same time based on convolutional neural networks. In order to achieve better performance, we extract discriminative nonlinear features that are invariant to pose and illumination. We propose to use the 1×1 convolutional kernels to extract the local features. Furthermore a parallel multi-stream convolutional neural network is developed to extract multi-hierarchy features which are more efficient than single-scale features. In the experiments we obtain the average face recognition rate of 96.9% on MultiPIE dataset. Even for profile position, the average recognition rate is also around 98.5% in different lighting conditions, which improves the state-of-the-art face recognition across poses and illumination by 7.5%.

1 Introduction

Face recognition has been one of the most active research topics in computer vision for more than three decades. With years of efforts, promising results have been achieved for automatic face recognition in both controlled [60] and uncontrolled environments [11, 17]. A number of algorithms have been developed for face recognition with wide variations in view and illumination, respectively. Yet few attempts have been made to tackle face recognition problems with the variations of pose and illumination [71]. In fact, face recognition is significantly affected by both pose and illumination which are often encountered in real-world images.

X. Jiang (✉) · Y. Hou · D. Zhang · X. Feng
School of Electronics and Information, Northwestern Polytechnical University, Xi'an, China
e-mail: xjiang@nwpu.edu.cn

X. Jiang et al. (eds.), *Deep Learning in Object Detection and Recognition*,
https://doi.org/10.1007/978-981-10-5152-4_3

Recognizing faces reliably across pose and illumination has been proved to be a much more difficult problem.

Pose can induce dramatic variations in face images. Essentially, this is caused by the complex 3D geometrical structure of the human head. The rigid rotation of head results in self-occlusion which means that some facial appearance will be invisible. At the same time, the shape and position of the visible part of facial images also vary nonlinearly from pose to pose. Consequently, the appearance diversity caused by pose is usually greater than that caused by identity. Thus general face recognition algorithms always fail when dealing with the images of different poses.

Illumination also can cause dramatic variations for face images. Assuming Lambertian reflectance, the intensity value $I(x, y)$ of every pixel in an image is the product of the incident lighting $L(x, y)$ and the reflectance $R(x, y)$ at that point as $I(x, y) = R(x, y) \times L(x, y)$. Thus, the captured images vary with the incident lighting. In order to achieve face recognition across illumination, there are two kinds of strategies. One is to extract illumination-invariant features from images, such as LBP [1] and HOG [65] et al.; the other is to model the distribution of illumination [24, 30].

In applications, both the pose and illumination variations exist. Thus a robust face recognition system should be able to deal with the two variations at the same time. Recently, the deep learning methods [27, 74] showed its great ability to model nonlinear distributions of data. It achieved the state-of-the-art performance in many fields of pattern recognition, such as object classification [58] and object detection [46]. Its great capacity is mainly due to the learning procedure which can find the hierarchical features from dataset. These features from each layer of the networks contain different levels of structure from a local gradient to its global shape. As a result, these learned features are more informative than traditional human-engineered features.

Even though different poses can induce the different appearances of the face, there exist some correlations between images of the same identity in different poses. Similarly, images of the same identity in different illuminations also correlate to each. Thus, through a proper learning method, the pose and illumination-invariant features can be obtained. Inspired by the excellent feature learning ability of deep convolutional neural networks, it is employed to develop an end-to-end face recognition method across pose and illumination in this work.

The remainder of this chapter is organized as follows: Sect. 2 briefly reviews the recent algorithms that deal with the pose problem. Section 3 introduces the algorithms that deal with the illumination variations. The proposed deep learning algorithm that can verify faces under pose and illumination variation is described in Sect. 4. The experimental results of the proposed algorithm are presented in Sect. 5. Finally, Sect. 6 concludes the chapter.

2 Pose-Invariant Face Recognition

Pose always causes substantial variations in the appearance of images due to the reason that images are the projection of 3D objects to 2D planar. Therefore when the pose of an object changes slightly, the appearance of the image will change dramatically. Consequently pose always brings difficulties for face recognition systems where the pose variation is unavoidable in uncontrolled environment. As a result, pose becomes one of the essential challenges for face recognition. Nowadays, researches also pay notable attention to deal with the pose variation problems. We can classify all the algorithms about pose variation into two categories: invariant representation-based algorithms and the model-based algorithms. For first category, invariant features or subspaces are constructed where the pose variation is removed, while the second type of algorithms tries to build up a generative model to predict the appearance of the object in different views.

2.1 *Invariant Representation*

In the classical frontal face recognition algorithms, face is always considered as a whole component. Therefore a lot of holistic approaches achieved quite good results. Principal component analysis (PCA) [62] is applied to find the eigenspace of face images; therefore face images can be represented by the projection values on those eigenvectors. Through the analysis, the dimension of face images has been reduced significantly, and the recognition is performed due to the distance in eigenspace. In fact, the assumption for these holistic approaches is that face position is fixed. Therefore, these algorithms try to find the relationship between corresponding pixel pairs among images. For the same person, corresponding pixels should have similar features, and the overall distance between images of the same subject is relatively smaller than that of different subjects. However, when the pose of the subject changes, the position of face components varies as well; consequently the correlation between corresponding pixel pairs is broken. Thus the holistic approach is no longer suitable for pose problem, but local components or features show their effectiveness in handling with the pose problem.

2.1.1 Engineering Designed Features

Landmarks (such as eyes, nose, and mouth) are the key points on face, which represent the key components in a face, as shown in Fig. 1. If the transform between the corresponding landmark points can be defined, then the same transform formula can be applied to convert two images. In order to find the same landmarks in images of different views, some robust feature extractors are used to describe the landmark points on faces.

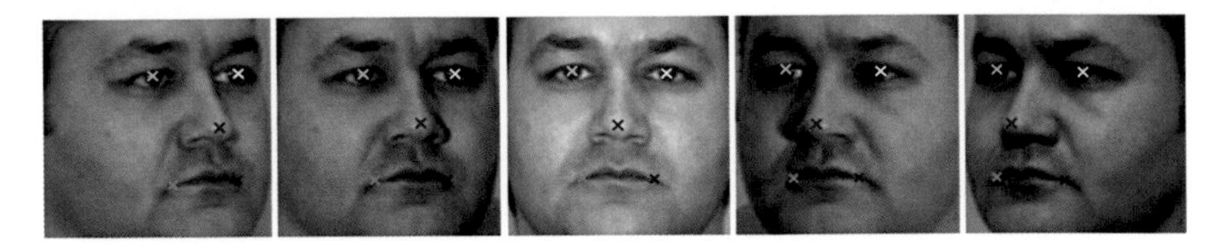

Fig. 1 The landmarks on facial images. Five landmarks are labeled in each image. The transform formula can be calculated from the relationship between the corresponding landmarks

Scale-invariant feature transform (SIFT) feature [37] is widely used in computer vision tasks for the extraction of robust features. It has also been used for face recognition. In order to find the connections of images for different poses, Biswas et al. [8] extracted SIFT features for landmark points, which can provide rotation and scale invariant features. Then tensor analysis was applied to learn the transform matrix between the landmarks of different poses. With this transformation, images taken in different views can be converted to the frontal view to compare with the frontal probing images for verification. Also, local binary pattern (LBP) is a descriptor that finds great success in texture analysis. It computes the distribution of local region variance and encodes the distribution into numbers, which are very efficient for further pattern analysis. LBP is also applied to extracted features for local regions around landmark points. Then all the local region features are connected into a new feature, which becomes pose-invariant [10].

In fact, accurate landmark detection itself is also a challenging problem. Besides extracting robust features around landmark points, researchers also tried to define some key points to find correspondence between images of different views. Dreuw et al. [6] propose to use speeded-up robust feature (SURF) to extract features in dense grid, and then RANSAC method is used to find the matching points between images of varied view for face recognition. Liao et al. [36] propose a partial face recognition method without alignment. First, they apply SIFT-like descriptors to extract key points from facial images. And then for those key points, sparse representation method is used to build up a complicated dictionary for all the possible local facial regions around key points based on training images. The key idea behind this method is also to extract robust features for key facial components but omit their locations.

Region-based pose-invariant feature extraction methods are also explored recently. Without the locations of key points, local regions are considered as the basic unit to contain key facial components. Ahonen et al. [2] propose to divide images into subregions, as shown in Fig. 2. Then they extract LBP features for each subregion of a face. Within a subregion, the location information is omitted; only the texture feature is extracted by the LBP descriptor. Thus the extracted feature is robust to pose variations as long as the key components are still located in the same subregion. It is reported that the proposed face recognition algorithm can keep good performance when the rotation angle is within 15°. For large pose variations, the content of each subregion changes greatly; consequently the correlation between subregions is broken.

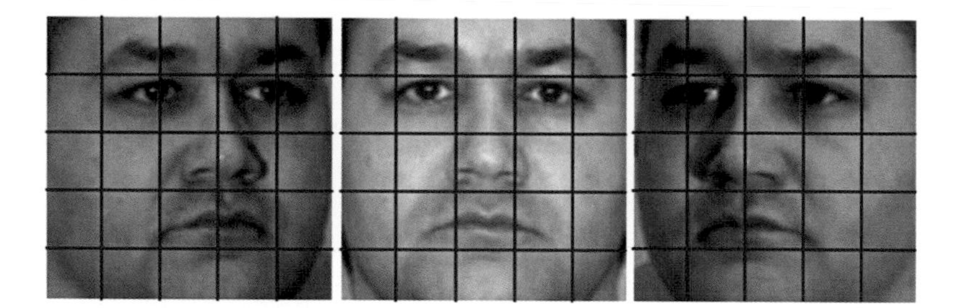

Fig. 2 Facial images are divided into local regions. Features are extracted based on each patch, thus the patch-based feature is invariant to pose

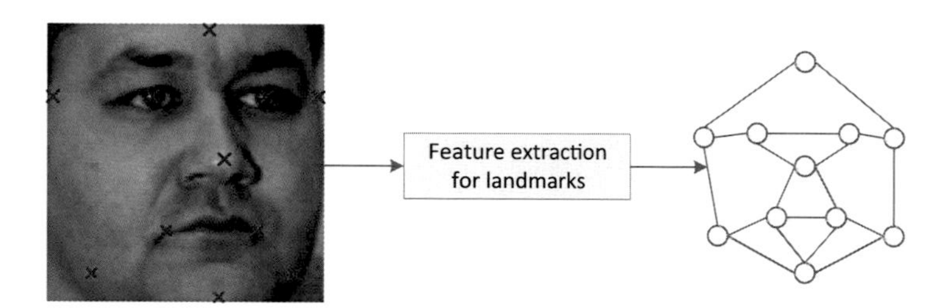

Fig. 3 Elastic graph for face recognition. The nodes of the graph are features extracted from local landmarks, and the edges of the graph represent the distance between neighboring nodes

Li et al. [33] propose a local region-based elastic matching method for face recognition across poses. Local descriptors, such as LBP or SIFT, are used to extract features for densely sampled subregions of images. The Gaussian mixture model is trained to extract the spatial-appearance distributions from the position of each local patch and its local feature. Each Gaussian model describes the relationship between corresponding patches of matched images. Then the verification is performed by a trained SVM that can discriminate the difference of Gaussian model between the matched and non-matched face images. In fact, the idea of elastic matching for face recognition of different poses is proposed by Wiskott et al. [67]. For each landmark of faces, a set of Gabor filters are applied to extract features, and then a graph with N nodes and E edges is constructed, where the nodes represent landmarks, and the edges are the feature distance between neighbored nodes, as shown in Fig. 3. Then the recognition is performed by comparing the graph of a probing face images to all the graphs of gallery images. The elastic bunch graph matching method can handle the rotation within 20°. For the elastic matching-based methods, a graph connects local components where the position of each component is also described by the graph. Thus the face can still be verified even though some local components are occluded.

Based on the cost of pixel-wised stereo matching, Castillo et al. [9] propose to do face recognition across poses. First they find three to four landmarks from face images to calculate the epipolar geometry parameters for gallery images. Then a stereo matching method is applied to find corresponding pixels between images.

Finally the cost of matching is used to identify the face images. Actually, there is an assumption that the corresponding pixels or components exist in two images. Thus when the pose changes greatly, the number of corresponding pixels between images decreases. Consequently, the performance of the algorithm drops significantly.

For the methods based on engineering-designed features, they try to find the corresponding local components between images. However, when poses change greatly, these manually designed features will always fail. The appearance of local components always varies greatly due to occlusion or out-of-plane rotation. Then the nonlinear correspondence should be found to describe the relationships.

2.1.2 Learning-Based Features

In order to find the nonlinear correspondence between images of different poses, some machine learning-based methods have been applied widely. Subspace learning methods are introduced to learn a new subspace that is invariant to pose variations. Metric learning methods are proposed to construct new distance measure methods, which are independent to the pose changes. Most recently, the deep neural network is also introduced to learn high-order nonlinear descriptors for images from different poses.

Linear Subspace Learning

In early years, principal component analysis (PCA) provides an important tool for extracting common features from dataset. The eigenvector that has the largest eigenvalue represents the direction of the biggest variance of the data, while the eigenvectors can be seen as the features shared among dataset. PCA-based methods achieve good performance in face recognition but are very sensitive to the misalignment of images. When face images are taken from different views, PCA encodes both identity and viewing conditions, which makes the performance of recognition degraded. Pentland et al. [42] propose to setup eigenspace for each component of the face, which only encodes the identity information, and the pose variance is alleviated by the selection of face components from images. When doing recognition, the reconstruction coefficients from each modular eigenspace are connected as a whole feature of the face.

Prince et al. [44] propose a statistical method to describe the distribution of face images regardless of pose. In the observed space, images from different views are located in different positions, where the difference caused by posture is much bigger than that caused by different identities. Thus it brings great difficulties for recognition. However, with the assumption that all faces of a single person in different poses can be described by a vector in the identity space, a linear transform mapping from the observation space to the identity space is proposed. Figure 4 shows the relationship between the two spaces. In the identity space, the pose

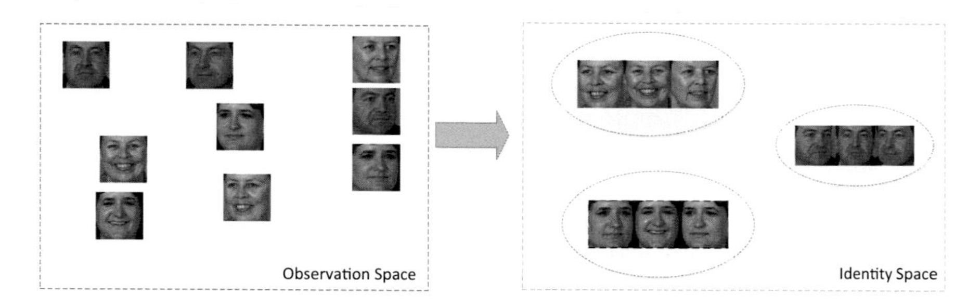

Fig. 4 Identity space only keeps the variance between different subjects but minimize the pose variance for a same subject

variance is diminished. Images from any pose can be represented as the linear combination of vectors in identity space h_i, and Gaussian noises ϵ_{ij} as follows:

$$X_{ij} = W_j h_i + \mu_i + \epsilon_{ij} \tag{1}$$

where W_j are the projection from identity space to the observation space and μ_i is the offset.

Therefore some generative models are introduced; the discriminative models are also used for dealing the pose problem actually. For discriminative models, they try to distinguish the difference between subjects regardless of pose variations. That is to maximize the margin between subjects or to find an optimal superplane to separate subjects.

Li et al. [31] propose to use canonical correlation analysis (CCA) to find a common space for images from different view, where the correlation between the same subjects is maximized, but not the traditional Euclidean distance. Correlation measures the difference of data tendency but not absolute distance. Thus it can allow some variance for data and becomes more robust to slight changes. The transform can be written as

$$\arg\max_{\omega_1,\omega_2} corr[\omega_1^T X_1,\ \omega_2^T X_1] \tag{2}$$

where $\|\omega_1\| = 1$, $\|\omega_2\| = 1$. X_i are images from different view and ω_i is the optimal transformation that can be solved by Lagrange multiplier method. In order to project images from all different poses to the same subspace but not only two views as CCA methods, Rupnik et al. [47] propose Multiview CCA(MCCA), as

$$\arg\max_{\omega_i,\ \dots,\ \omega_k} \sum_{i \neq j} corr[\omega_i^T X_i,\ \omega_j^T X_j] \tag{3}$$

where $\|\omega_i\| = 1, i = 1, \dots, k$. The set of transform ω_i can transform images from different views to the same subspace and meanwhile keep the maximal correlation for the images of the same person. Based on MCCA, Sharma et al. [53] improve the algorithm further. They first find the optimal mapping functions for images from different views with the MCCA method, and then linear discriminative analysis (LDA) is applied to classify each subject in the new subspace.

Sharma et al. [52] propose a unified framework for multiview analysis called generalized multiview LDA (GMLDA), in addition. Within this framework, the discriminative analysis for each subject and the self-correlation of the images from the same subject are combined as

$$\arg\max_{\omega_i} \sum_i \mu_i \omega_i^T S_{bi} \omega_i + \sum_{i \neq j} \lambda_{ij} \omega_i^T Z_i Z_j \omega_i \tag{4}$$

where $\sum_i \gamma_i \omega_i^T S_{wi} \omega_i = 1$, μ_i, λ_{ij} and γ_i are parameters for linear combination. S_{bi} and S_{wi} are the between-class and within-class scatter matrix for the ith pose. The first term performs the LDA analysis for different poses, which enhances the distinguishability. On the other hand, the second term focuses on the correlation for all the images from the same subject. That is, the projection from different poses should be close to each other in the latent space. Z_i are the matrices whose column contains images of the same subject.

For CCA methods, it requires each subject to have exactly the same training data for each poses, while GMLDA only requires pairwise training data from different poses. In order to alleviate the requirement for training data, Kan et al. [28] proposed multiview discriminant analysis (MvDA). They apply the idea of LDA to analyze pose problem, where the intrapose scatter is minimized and interpose scatter is maximized. Then all the images from the same pose will be cluttered together in the new subspace.

Nonlinear Subspace Learning

The pose variations for images are due to the projection of 3D structure of the object to the 2D planar. As a result, poses actually bring nonlinear transform for the appearance of images. Consequently the nonlinear models should be more suitable than linear models for the description of pose variations. Kernel-based methods are the direct extension of linear methods, where the kernel can transform a linear subspace to a nonlinear subspace. In a nonlinear subspace, the nonlinear distributed data can be separated by a linear surface. Consequently, the classification can be achieved by linear methods in a higher-dimensional subspace.

There are quite a few adaptations to the linear methods, such as the kernel-PCA [49] is the extension of PCA by kernel method. Yang et al. [69] proposed a kernel Fisher discriminant framework by full usage of the KPCA and LDA. Experiments show the improvement in face recognition tasks. Recently, Sharma et al. [55] proposed a generalized multiview analysis (GMA) method which projects a pair of images from different views into a common space by using the kernel tricks. The proposed method shows its effectiveness in pose-invariant face recognition.

Metric Learning

Besides using a suitable subspace to represent all the images from different poses, the distance metric also can be adapted to deal with pose variations. Schroff et al. [51] propose to compare the similarity of probe images with a big set of gallery images, and the similarity list is used as a feature to determine the identify of probe images. It is based on the assumption that images from the same person should have more common look-alike samples than that from different people, even if the images are taken in different conditions. Liao et al. [35] propose to compare the low-frequency information of the probe image with all the gallery images, and then a pooling method is applied to make it pose-invariant. Furthermore, Kafai et al. [25] construct reference face graph (RFG) to represent the relationship between different subjects, where each node in the graph contains all the images taken in different conditions of that subject, as shown in Fig. 5. The importance of each node is readjusted by its node centrality including degree, betweenness, and closeness for weighted graphs. Finally, the probe image is represented by the vectors that are composed of the similarity measure to each node in the graph where the hashing code is calculated for each oversampled region of images.

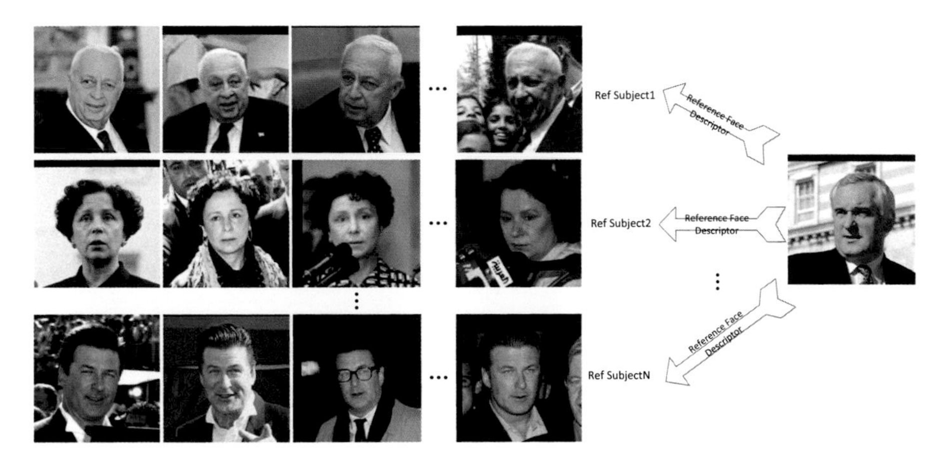

Fig. 5 In the reference face graph, each node is composed of faces from different views. Then all the faces are used as the basis to represent the probe face where the reference face descriptors are calculated

Deep Learning

Recently, deep learning-based methods show great success in the field of signal processing. It achieves the state-of-the-art performance in many applications such as face recognition, object classification, and so on. The great ability behind the neural networks is the nonlinear modeling capability actually. The nonlinearity is due to the nonlinear activation neurons in the networks. In addition, the multiple layers of the neural network make the order of the nonlinear model much higher than traditional models, which can represent the data more accurately.

Andrew et al. [3] proposed to use two parallel neural networks for the feature extraction of images from different poses, and then the output items from the networks are maximally correlated, where the correlation value is used to optimize the parameters of the neural networks. In fact, deep networks are performed as nonlinear mapping functions for input images. As a result, the deep canonical correlation analysis (DCCA) method shows better performance than Kernel CCA and CCA in the experiments, which is due to the robust features extracted by neural networks. The structure of the DCCA is shown in Fig. 6, where a three-layer fully connected neural network is applied. With the improvement of the neural network, more sophisticated features can be extracted.

Zhu et al. [73] propose a deep network to find the identity preserving features from images in different views, as shown in Fig. 7. There are three convolutional layers in the deep network. The input of the network can be images from any

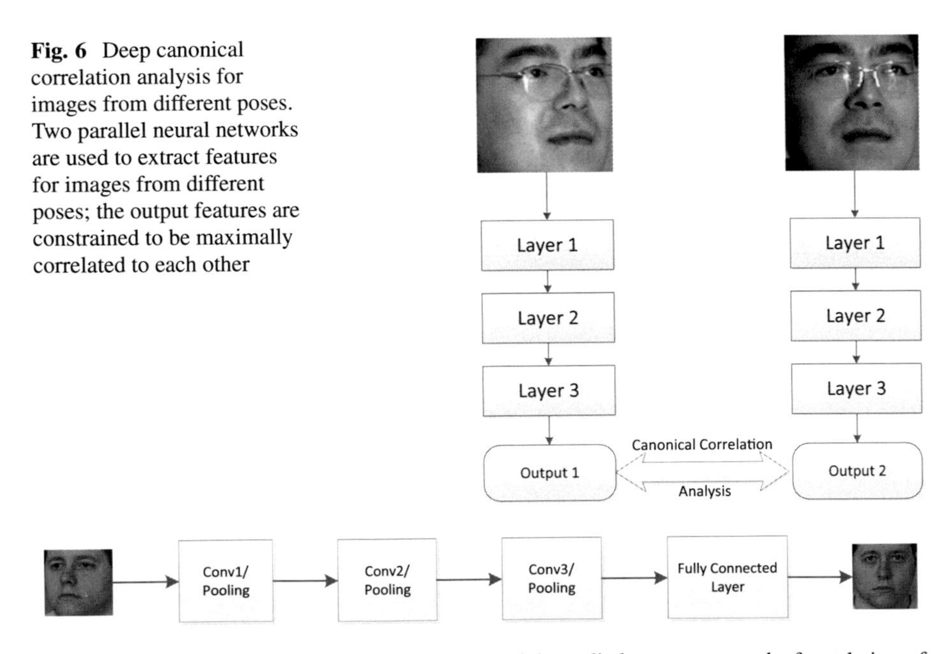

Fig. 6 Deep canonical correlation analysis for images from different poses. Two parallel neural networks are used to extract features for images from different poses; the output features are constrained to be maximally correlated to each other

Fig. 7 Identity preserve network. The neural network is applied to reconstruct the frontal view of input images from different poses

poses, while the output of the network is the reconstruction of the frontal view for the input subject. The network is composed of two basic units, which are the feature extractor and the reconstructor for the frontal view. In training, the difference between the reconstructed images and the ground truth is calculated for back propagation. Through the supervised learning procedure, the network can recover any input image to its corresponding frontal view. The extracted features also show great capability for identity discrimination. The face recognition is performed by comparing the recovered frontal face with the gallery frontal images. For this network, the last reconstruction layer is achieved by fully connection layer, which has millions of parameters needed to be trained. Consequently, it requires a huge number of images for training.

Based on the encoder framework, Zhang et al. [70] propose an encoder network for images of different poses. There is only one hidden layer for this network. The input of this network is images from different poses, and the output is the frontal image of the same subject. The encoder tries to find common features for images from different views. Furthermore, they propose to use random images to represent identity of each subject and train the encoder to find discriminative features for each subject. In order to keep the convergence of training, the sparse constraint of parameters is added to the loss function. Figure 8 shows the structure of the network. Compared with Zhu's work [73], it only has one hidden layer, which reduces the number of parameter but also reduces the capability of model. It also proves that increasing the layers of neural network can enhance the order of nonlinearity of the model, which can increase the discriminability of the model. Kan et al. [26] also noticed that only one hidden layer is not enough to model the nonlinear transform from any pose to frontal view. Thus they propose to use a cascade of autoencoders to transform the pose gradually from non-frontal view to frontal view, as shown

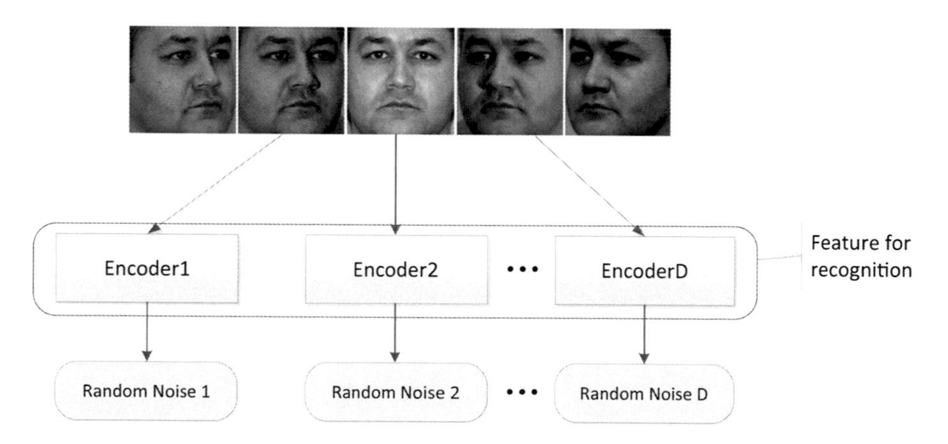

Fig. 8 Random face learning network. The random images are used as the output of one-layer encoder to learn the features for images from different views. Then the learned features are used for recognition

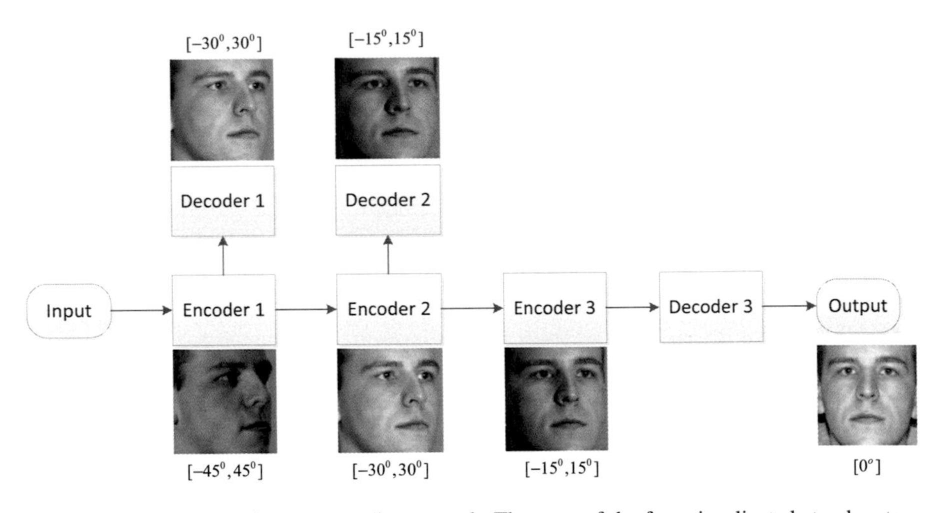

Fig. 9 Stacked progressive autoencoder network. The pose of the faces is adjusted step by step, and then all the trained encoders are stacked to compose a network to recover profile images to frontal ones

in Fig. 9. Altogether, three encoders are stacked together, which can reduce the probability of being trapped into local minimal during training.

For all these autoencoder-based methods, the key idea is to reconstruct the frontal face from non-frontal input. Then the recognition is performed based on the reconstructed images. Always, these methods separate the recognition tasks into two independent steps, which is not an end-to-end procedure.

Kan et al. [27] propose a two-stage network for face recognition across poses, as shown in Fig. 10. For the first stage, images from different views are input into different sub-networks for the extraction of view-specific features, and then all the features are fed into a common sub-network for the extraction of common features across poses. In addition, the network is trained based on the Fisher principle, where the intra-view distance is minimized and the inter-view discrepancy is maximized. With the trained network, the features from topmost layers are used for classification. For MvDN, it first requires the input images to be classified into groups due to poses, and then images can be fed into the proper network. In addition, it is not an end-to-end framework for the task of recognition.

Majumdar et al. [38] propose to use autoencoder for the extraction of image features. In order to make the feature more robust to pose, a whole face image is decomposed into several local patches that contain the main components. Also, the sparsity constraint is applied to the autoencoder. For classification tasks, the input images are fed to the autoencoder for feature extraction, and then a classifier is applied for verification. Even though the patch-based method can improve the robustness for pose, it requires accurate segmentation results from the preprocessing methods.

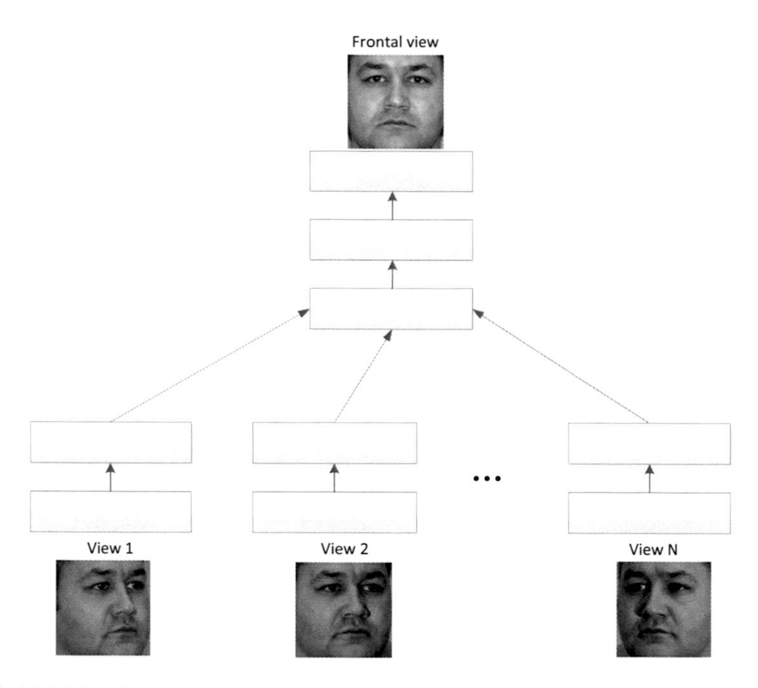

Fig. 10 Multiview deep network. The network first extracts features for different views separately, and then the features are integrated in the following network. In training the Fisher distance between the frontal view and different views is used as the error to optimize the parameters

Recently, Peng et al. [41] propose a deep network to extract pose-invariant features. First they use synthesis methods to enrich the training samples, and then identity and non-identity features are extracted through a multitask learning procedure. Finally, the pose-invariant features are purified through the constraint of reconstruction errors from different poses. The proposed method requires the synthesis of non-frontal faces for training. Also, it requires the pose and landmark labels for each training image, which is not easy to get in the applications.

Masi et al. [39] realized that the frontalization of non-frontal faces is actually very challenging and becomes harder with the increasing of rotation angle. Actually, it is a highly nonlinear transform, and many corresponding information between frontal and profile faces is lost. Thus they propose to develop separated network for different poses, called pose-aware CNN models (PAM). That is, for different poses, e.g., frontal, half-profile, and profile images, different CNNs are trained. Then averaging the scores from all different PAMs gives the recognition results. For this method, it requires the input images to be rendered into different poses, and then they can be fed into corresponding PAMs for recognition. Even though they constrain the range of rendered image, it is still a challenge to produce images of different views.

Tran et al. [61] introduce generative adversarial network (GAN) for the task of pose-invariant face recognition. The GAN network is constructed based on the

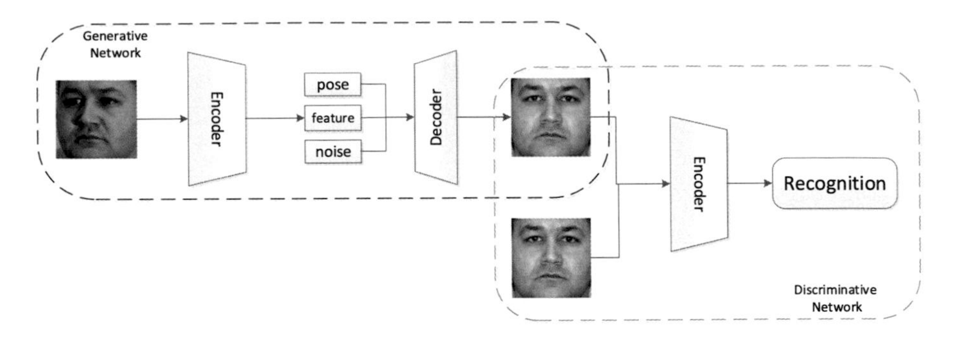

Fig. 11 Generative adversarial network (GAN)-based pose-invariant face recognition. In the generative network, the encoder-decoder structure is used. Also, some side information is introduced for the extraction of robust features in the network

framework of encoder and decoder. Through the generative network, face images from any poses are converted to frontal ones. The discriminative network is used for the judgment of the new created images. In order to enhance the extraction of pose-invariant feature in the generative network, pose label and noises are used as side information. The structure of the proposed network is shown in Fig. 11. Even though GAN network is more powerful than traditional neural network, the convergence of the network is still a great limitation for its application.

Neural network provides a powerful tool to extract features from training images, which has been applied to solve the pose problem. However, how to design a suitable network structure for this specific problem is still an open problem for researchers.

2.2 *Synthesis-Based Methods*

Pose variations introduce nonlinear transform for images of the same subject. Besides pose-invariant features that can be used as a low-dimensional representation for the image, researchers also tried to synthesis the frontal face images directly from images of arbitrary poses. The synthesis-based methods can be further classified into the 2D-based and 3D-based methods, depending on if the 3D model of face is applied or not.

2.2.1 2D-Based Synthesis Methods

2D-based synthesis methods try to convert images of varied poses directly to the frontal faces. Based on the different units to process, all the 2D synthesis methods can be classified into three categories: triangle mesh wrapping, patch wrapping, and pixel wrapping.

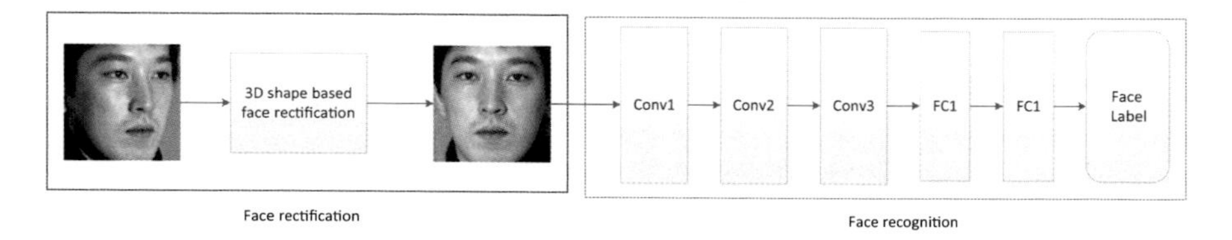

Fig. 12 Profile faces are first rectified to frontal face based on triangular mesh, and then deep network is applied to recognize faces in the frontal view

In the early days, a 3D model of a subject is usually presented by a triangle mesh. Consequently, the triangle mesh can also be casted on 2D images, and each triangle is used as a unit to calculate the transform between two different poses. Taigman et al. [59] apply triangle mesh for alignment of the input images. After all the input images are converted to frontal face, a deep network is applied for feature extraction and recognition of the face image, as shown in Fig. 12. For the alignment, key landmarks are detected and then triangular mesh is cast to the 2D image, where the 3D shape of the face is correlated with the triangular mesh. Finally the frontal facial image can be estimated due to the affine wrapping of triangular mesh. With accurate adaption of input images to frontal view, the following neural network can perform face recognition under varied poses. For mesh-based wrapping, it mainly depends on how well the triangle mesh can be cast to the 2D face image.

In order to avoid the detection of landmarks on face, Ashraf et al. [4] propose to decompose images into small patches, and for each patch, a learned transform can be applied to convert the non-frontal patch to be frontal ones. They treat each patch as independent unit for the whole procedure; however, there are close connections between neighboring patches. Thus those patches are highly correlated to each other where the relationship can be applied in solving the pose problem. Recently, Ho et al. [19] consider this relationship. The searching for optimal transform for each patches are converted into an optimization problem, where the reconstruction error of each patches should be small; meanwhile the transform for neighboring patches should be similar. The additional items constrain the smoothness of the global transforms.

For patch-based synthesis methods, they treat a local patch as a unit to calculate the transform. However, the nonlinear transform between poses is different from pixel to pixel. Thus Li et al. [34] propose a pixel-based transform. They learn a set of template displacement model from 3D dataset first. Then for each input image, the template displacement model is applied to transfer the images of arbitrary poses into frontal face pixel-wisely. For the occluded part of the face, the information is compensated from frontal face. Even though the proposed method can reconstruct the frontal face, only the non-occluded region will be used for verification.

2.2.2 3D-Based Synthesis Methods

For 2D-based wrapping methods, they directly find the transform between different poses. When the pose changes significantly, the mapping between different poses becomes highly nonlinear. As a result, the 2D wrapping results will become worse. In fact, pose problem is caused by the projection of 3D subject model to 2D imaginary surface. That is, the intrinsic 3D model controls the appearance of the images. Thus researches also try to use 3D model for solving pose-invariant recognition problem.

Ding et al. [12] introduce a 3D model-based dense-mapping method for the recovery of frontal face. First, the key landmarks are detected from the 2D images of arbitrary pose, and then they are matched to the corresponding landmarks in a standard 3D face model. As a result, the pose transform can be estimated for the input 2D image. In order to recover the texture, a dense mapping is used with the estimated pose transformation matrix. Furthermore, homography-based patch correction method is proposed to enhance the realisticity of the recovered texture. If there is occlusion in the original 2D facial images, then the recognition is only based on recovered un-occluded part.

Further, Ding et al. [13] transform the profiled face image recognition problem to the partial face recognition problem. Based on the key point mapping between 2D images and 3D face model, the profiled images are transformed to images of frontal view. Then sparse coding-based feature is extracted on the reliable regions of the recovered frontal view.

3 Illumination-Invariant Face Recognition

Illumination is another big challenge for face recognition. The intensity value of each pixel I_{xy} in an image is determined by the strength of the incident light, L_{xy}, and the angle of the incident light θ_{xy} and the reflectance rate of the surface R_{xy}, as $I_{xy} = \oint L_{xy} R_{xy} \cos\theta_{xy} d\Omega$. Thus, when the incident light changes, the appearance of the same object will vary as well. Always, the variation that is caused by illumination is more significant than that of subject. Consequently, lighting always causes the degradation of face recognition methods. Algorithms that try to remove or alleviate the lighting variations can be classified into three categories as image processing-based methods, invariant feature-based methods, and illumination model-based methods.

3.1 Image Processing-Based Methods

Lighting is one of the factors that control the appearance of images. The average intensity value of images that are taken in brighter situation is bigger than that in dark situation. Thus researches proposed to use image-processing methods to enhance the intensity value of images, such as histogram equalization (HE) and gamma correction.

Histogram equalization [43] tries to adjust the histogram of input images. That is to adjust the intensity distributions of images. In fact, the pixel intensity value of an image that is taken in brighter situations is bigger than that of a darker image. Thus the intensity distribution of brighter images will have peaks in bigger value region, while that of darker images will have peaks in smaller value region. Using histogram equalization will make the distribution of intensity value evenly. That is, the brighter images will become dimmer, and darker ones will become brighter. Histogram equalization only considers the intensity value of each pixel but not the sematic meaning. Therefore it will always introduce some abrupt noises in images due to the assignment of new intensity value to all the pixels of the same intensity value, as shown in Fig. 13b.

Gamma correction is the more dedicated adjustment for the intensity distribution, while histogram equalization turns the original distributions into uniform distribution. Gamma correction function is defined as

$$T(I) = I_{\max}\left(\frac{I}{I_{\max}}\right)^{\gamma} \tag{5}$$

where I is the intensity value of current pixel. I_{max} is the maximum intensity value in the image, and γ determines the curve for adaption, which is the key parameter. According to different gamma curves, the original intensity values can be modified to any desired distributions. Normally the darker pixels are tuned to be brighter, while the bright pixels are kept, as shown in Fig. 13c.

Huang et al. [21] proposed an adaptive gamma correction method, where γ is determined by the cumulative distribution of intensity, $cdf(I)$, as

$$T(I) = I_{\max}\left(\frac{I}{I_{\max}}\right)^{1-cdf(I)} \tag{6}$$

$$cdf(I) = \sum_{I=0}^{I_{\max}} \frac{pdf(I)}{\sum pdf} \tag{7}$$

where $pdf(I)$ is the probability density function of intensity. The proposed method combines gamma correction and histogram modification method. Similar to the histogram equalization, gamma correction is also a holistic modification method, which does not consider the local information in the image. Jiang et al. [23] proposed to combine the local and global information for the lighting augment. The local factor I_a^{local} provides the local variance, while the global factor I_a^{global} provides the overall intensity of the image. These two kinds of information are combined by a bilinear method, and then a perception-based method is used to adjust the brightness of the images, as

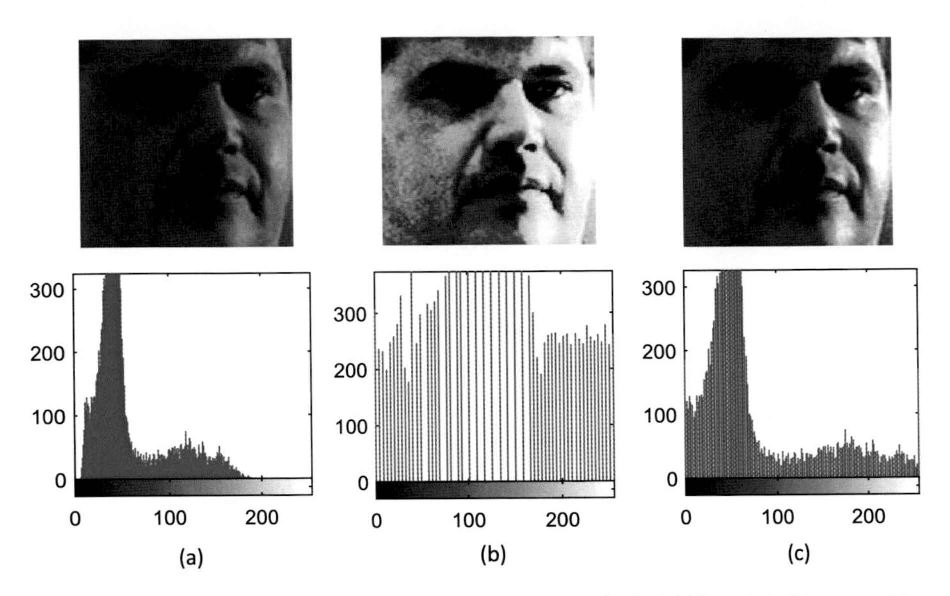

Fig. 13 Comparison of two basic illumination adjustment methods. (**a**) The original image and its histogram. There are two peaks in the histogram due to the side light. (**b**) Histogram equalization result. The histogram is adjusted to be equally distributed but leaves many noises in the image. (**c**) Gamma correction result. Compared with histogram equalization result, there is few noise introduced into the result image

$$Y(\alpha, m, f; I) = \frac{I}{I + (f I_a)^m} I_{\max} \tag{8}$$

$$I_a = \alpha I_a^{\text{local}} + (1 - \alpha) I_a^{\text{global}} \tag{9}$$

where α, m, and f are parameters determine the detail, contrast, and brightness of the image. The model is derived from human vision perception system. Image processing-based illumination adjustment methods mainly focus on the modification of the intensity distribution and aim to brighten dimmed images. This kind of method only considers the appearance of current images, but not the factors that cause the current appearance. Thus these methods always cannot solve the lighting problems thoroughly.

3.2 *Invariant Feature-Based Methods*

Images are the cooperative results of illumination and objects. Even though the illumination can vary due to different situations, objects themselves do not change. Therefore researchers try to find illumination-invariant representations from images to describe the intrinsic features of objects. Edges describe the shape or contour

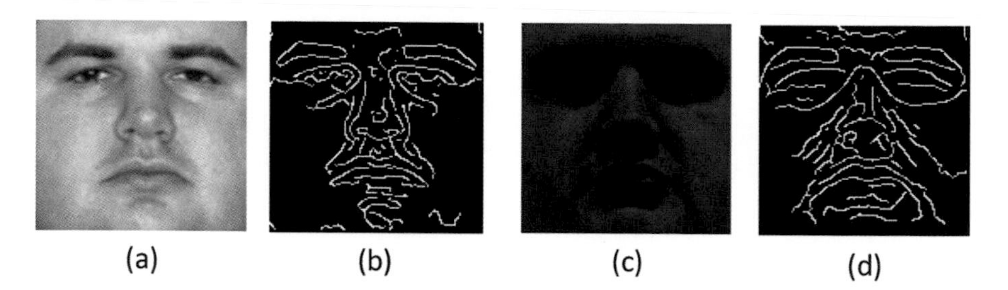

Fig. 14 Edges are used as features for face recognition. (**a**) and (**c**) are the original images, (**b**) and (**d**) are the edges for (**a**) and (**c**), respectively. Compared to (**b**) and (**d**), the edges are different for the same person under different lighting conditions. Thus the edge features are not absolutely invariant to illumination

of objects, which are considered as one of the illumination-invariant features. On the contrary, color of the object will vary according to its situation. Gao et al. [15] propose to use edges on the face image to perform face recognition. All the face components, such as eyes, nose, mouse, and eyebrows, are represented by line contours, as shown in Fig. 14. Then the distance between lines is calculated for verification. Zhou et al. [72] apply multi-scale Gabor filters to extract features from face images, where the multi-scale edge features are extracted. Some other popular feature descriptors such as local binary pattern (LBP) and scale-invariant feature transform (SIFT), which extract edge-based features for local regions, are also considered to be robust to illumination. However, shadow will also produce edges, even obvious edges in images, which are quite difficult to be distinguished from edges of the object. Consequently, these edge-based features can only work well with slight lighting variations.

With the simplified illumination model $I = L \times R$, Shashua et al. [56] propose the concept of quotient images, which is the ratio between a testing image I_y and a linear combination of three images I_j with weight x_j, as

$$Q_y = \frac{I_y}{\sum x_j I_j} \tag{10}$$

where the combined lighting condition of I_j is similar to the lighting condition of I_y. Thus the quotient image only relates to the reflectance of the object and is free from the lighting variations. With quotient image Q_y, images under new lighting condition can be rendered and furthermore can be used for the face recognition in different lighting conditions. However, quotient image is based on the assumption that the same class of object all has the same shape. It is a very rough assumption. In fact, every face is different. Wang et al. [64] extend the concept of quotient images. They propose to estimate the lighting map from images directly. According to the Retinex theory [60], most lighting information can be considered as the low-frequency signal, and most reflectance information is high frequency. Thus lighting information can be estimated from the low-frequency part of the original images. The proposed self-quotient image is defined as

$$Q_{sy} = \frac{I_y}{F * I_y} \tag{11}$$

where F is a smoothing kernel.

Also based on the theory of Retinex, Xie et al. [68] propose a two-step strategy for the normalization of lighting conditions. First, they decompose the input images into the low-frequency and high-frequency parts using the total variation model. Then the two decomposed components are normalized, respectively. The normalization of the low-frequency part will enhance the uniformity of lighting conditions. Then the normalized high-frequency and low-frequency parts are multiplied together to get the normalized images. In the second step, kernel eigenspace is used to correct the visual flaws of the normalized face images. Even though KPCA can be used to improve the appearance of the image, it requires a lot of training images of each subject for the construction of kernel subspace.

He et al. [18] realize that the distribution of face subspace is a nonlinear manifold; thus, the nonlinear method should be more suitable for the problem. They propose to find the face manifold based on the locality preserving projection, which is called as the Laplacian face representation. This subspace can keep the identity difference but minimize the other variance within a same subject.

Compressive sensing theory provides a dramatically new method to represent signals. Based on the theory, continuous signals can be sampled randomly which breaks through the constraint of Shannon theorem. From the training dataset, a sparse representation of the subject can be learned with the sparse constraint. Wagner et al. [63] construct a sparse coding dictionary from a set of images taken in different lighting conditions and different poses. Images in various conditions are recovered to classical frontal images of the same subject with the sparsity constraint. Then the learned sparse representation is invariant to illumination. However, the sparse dictionary learning requires images from all different conditions to keep the performance of the proposed algorithm.

Recently, deep learning-based methods are also applied to extract illumination-invariant features for images under different lighting conditions. The classical deep learning methods, such as AlexNet [29] and VGG-Face network [40], extract features from the convolutional layers, and then discriminative features are classified by the fully connected layers. Besides the classical neural network structures, researchers also adapt loss functions to improve the extracted features [20, 66]. These methods do not focus on the lighting problems but try to extract robust features for general face recognition problem. Therefore the structure of the network is not specifically designed for lighting problem.

3.3 Illumination Model-Based Method

Illumination is an essential factor for imaging. Therefore, researchers also try to analyze the distributions of images that are taken in different lighting conditions.

With the illumination model, images under different lighting conditions can be reconstructed. Also, the illumination can be changed or removed from the images. Belhumeur and Kriegman [7] introduce the theory of illumination cone, which is the basic theory for the lighting space of an object. If an object has convex shape and Lambertian surface, then all the images about this object can form a polyhedral cone. The dimension of this cone is determined by the number of distinct surface normal vectors of the subject. In practice, the illumination cone theory can be relaxed to objects of any shape and with a general reflectance surface. The authors also point out that the illumination cone of an object could be approximated by a low-dimensional subspace. The illumination cone theory only illustrates the structure of lighting space for an object in a certain pose. The relationship between illumination cones for different poses is still unclear.

In fact, a completed high-dimensional illumination cone is always difficult to build in practice. Thus, researchers try to find the low-dimensional approximation for an illumination cone. The illumination effect for an object can be considered as the convolution of incident illumination with the reflectance function of the object. Given the 3D model of an object, the lighting subspace can be constructed by spherical harmonic basis [5, 45]. All the bases are given with implicit equations, which are functions about illumination and object surface normal vectors. Given the 3D shape of the object, lighting position, and intensity, it is very easy to obtain the basis for the lighting subspace directly. With the first three orders of the harmonic basis, 90% of the illumination effect can be estimated, as shown in Fig. 15. However, the requirement of deep information for the object also limits the application of the method.

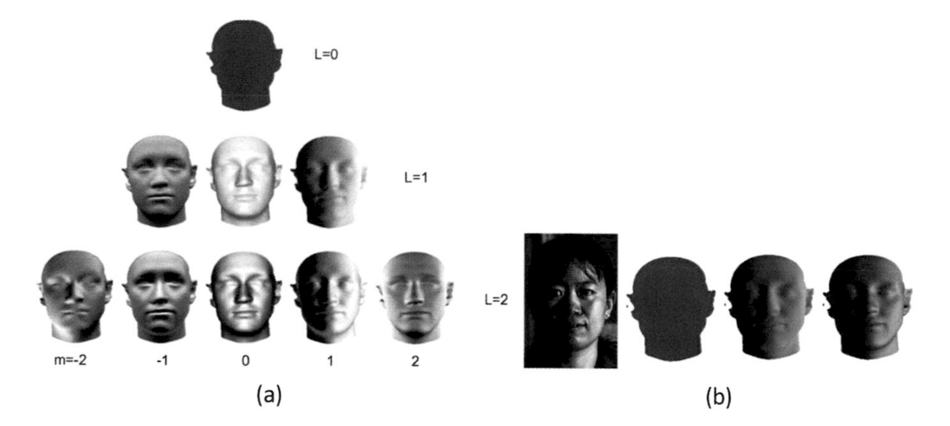

Fig. 15 Lighting map estimation based on spherical harmonics basis. (**a**) the first three orders of the spherical harmonic basis of a face. Each row is the basis of the same order. From top to bottom, they are the basis of order 0 to order 2, respectively. (**b**) From left to right, they are the original face image, lighting map reconstructed by the basis of order 0 to 2, respectively. (Reprinted from Ref. [22], with permission from Elsevier)

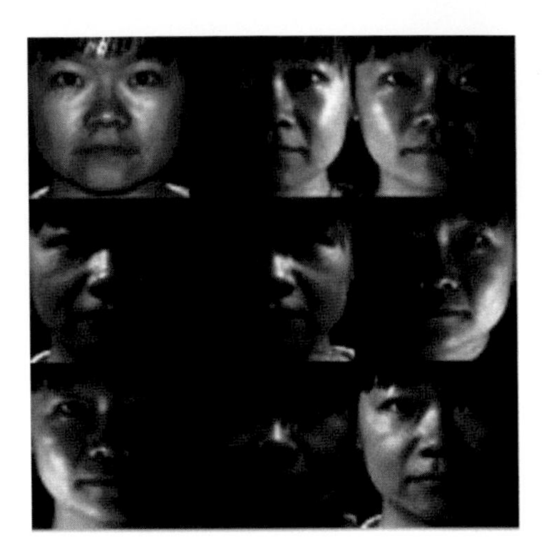

Fig. 16 Images taken in nine specific lighting conditions can be used as the basis to construct the lighting subspace of the object. (Reprinted from Ref. [24], with permission of Springer)

In order to avoid the requirement of 3D model of subject or the learning procedure to find lighting subspace, Lee et al. [30] propose to construct the low-dimensional approximation directly from real images. Those real images are found through minimizing the distance between two subspaces, where one is the spherical harmonic subspace H and the other is the selected real image subspace, C. Then the real lighting configuration of those selected images can be used for any subject to construct the low-dimensional lighting subspace, C. That is, the real images taken under those lighting configuration can be used as the basis images for its lighting subspace. This paper proves that the lighting subspace constructed from real images is a very good approximation of spherical harmonic subspace. In application, only nine lighting positions are required to build up the lighting subspace, as shown in Fig. 16. Also, it does not require the depth information as the traditional spherical harmonic subspace. However, the specific lighting configuration sometimes cannot be accessible. In real application, there are always a few sample images or even one image of each subject.

Considering the practical application of lighting subspace estimation, Jiang et al. [24] propose to create the basis images from any sample images of an object. That is, given one image that can be taken in arbitrary condition, the nine basis images can be reconstructed, and consequently the lighting subspace can be set up, as shown in Fig. 17. In this paper, the estimation of basis images is based on the maximum a posterior estimation. The basis images are composed of the common components and personal components, where the common component is the mean value from the training images and the personal components describe the specific characteristics of each subject. Thus the estimated basis images can recover both the shared and the individual features for each subject. This method breaks the requirement of nine real images that are taken in specific lighting condition.

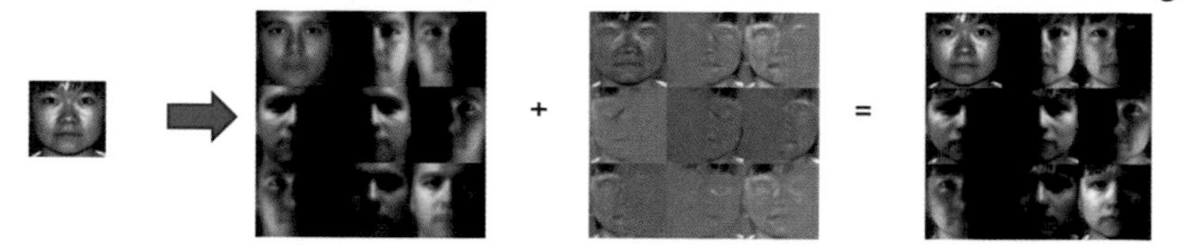

Fig. 17 The basis images of lighting subspace can be estimated from images under arbitrary lighting conditions

4 Multi-stream Convolutional Neural Networks

In general conditions, the illumination of the environment and the pose of the object are always uncontrolled. Therefore the robust face recognition system should be able to process the pose and lighting problems at the same time. The current algorithms that are dealt with pose and lighting problems are introduced in Sects. 2 and 3, respectively. Besides these specific designed algorithms, there are also some methods doing face recognition in general conditions. Especially with the development of deep learning methods, some deep neural networks are designed for the face recognition problems. Schroff et al. [50] propose to enhance the discrimination of the deep features according to the standard that the distance within a class is minimized and the distance between classes are maximized. Sun et al. [57] increase the dimension of the hidden layers and add constraint for early convolutional layer to increase the discriminative power of the neural network. The proposed network is called as DeepID2+, which improve the performance for the face recognition in natural conditions.

For face recognition across pose and illumination, the global structure of images is destructed by views; meanwhile, lighting brings wide variations for the appearance of images. Thus the pose- and illumination-invariant features should be local but not global. Furthermore the multiple hierarchical features are always much more informative than features in a single scale. Consequently, we propose an end-to-end convolutional network which can extract multi-hierarchy local features for the task of face recognition. The overall architecture is shown in Fig. 18. In our proposed networks, the input is a facial image under an arbitrary pose and illumination. The output is the identity label for the face image.

4.1 Root Convolutional Layer

Recently convolutional neural networks (CNN) show great performance in different fields of computer vision, such as object detection [46] and object classification or recognition [58]. The superb capability of CNN is mainly due to its high-order

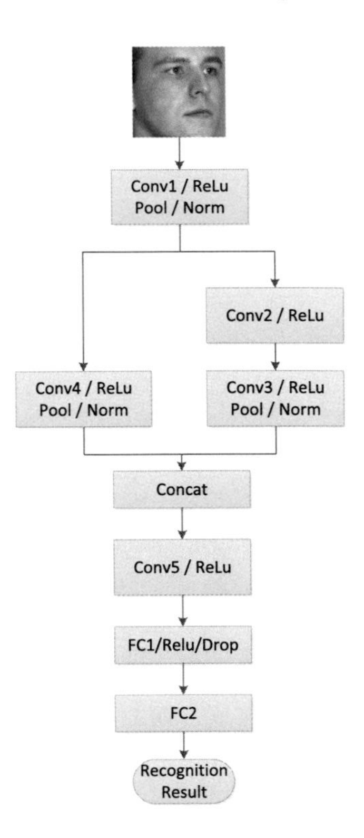

Fig. 18 Architecture of the proposed deep network. Conv1 has the kernel size of 11×11 and the dimension of 96. Conv2, Conv3, Conv4, and Conv5 all have the kernel size of 1×1 and the dimensions of 200, 400, 300, and 500, respectively

nonlinear representation for data. In practice, CNN extracts features from input images layer by layer through convolution kernels. For the proposed networks, the input images, x_i^{l-1}, are cropped and mirrored to the size of $w \times h \times c = 227 \times 227 \times 3$. Then they are fed into a convolutional layer (Conv1) k_{ij}^l with 96 filters of size $11 \times 11 \times 3$. The output, x_j^l, of this convolutional layer is written as

$$x_j^l = \sum_{i \in M_j} x_i^{l-1} * k_{ij}^l + b_j^l \tag{12}$$

where l is the layer index, b_j^l is the additive bias term, and $*$ represents convolution in a local region M_j of input signals. In this convolutional layer, 96 filters are applied locally to the whole images resulting in a feature map of size $55 \times 55 \times 96$. Then rectified linear unit (ReLU) is applied to the extracted feature map. ReLU serves as an activation unit in the network which brings the nonlinearity to the feature. Here we use a ramp function $f(x) = max(0, x)$ to rectify the feature map. This activation function is considered to be more biologically plausible than the widely used logistic sigmoid or hyperbolic tangent function.

Consequently, the rectified feature maps will be given to a max-pooling layer (Pool) which takes the max over 3×3 spatial neighborhoods with a stride of two for each channel, respectively. Through max-pooling operation, features will become insensitive to local shift, i.e., invariant to location. Afterward, those features will go through the local response normalization layer (Norm), which performs the lateral inhibition by normalizing over local input regions. In our model, the local regions extend across nearby channels but have no spatial extent. For normalization, each input value is divided by the sum of local region as shown in Eq. 13:

$$s(x_i) = (k + (\alpha/n)\sum_i x_i^2)^\beta \tag{13}$$

where n is the size of each local region and the sum is taken over the region centered at that value x_i (zero padding is added where necessary). From the root layer, we can obtain the local feature set which mainly contains all kinds of edges in different orientations. Generally edges are considered as illumination-invariant features. Since local structures are more important in our case, we will continue to seek for local features instead of global features which are normally obtained in further layers of traditional CNN.

4.2 Multi-hierarchical Local Feature

In fact, the window size of the convolution kernel is considered as the receptive field for feature extraction. That is, bigger windows can include information in wider range for processing. For the face images from different views, the global structures of images change diversely. However, there is a tight correlation among local regions of images taken in different views. Therefore the pose-invariant features should be local features, and in addition the spatial information should be kept for each local feature. Accordingly, smaller windows should be applied to extract features. Here we propose to use the kernel of size $1 \times 1 \times c$. With the kernel of 1×1, no spatial patterns across multiple pixels are extract, but the patterns between c channels are learned without losing the location information for each pattern. Thus the feature can keep the correlation among different views. Meanwhile, the number of parameters will also be reduced for the 1×1 kernel size compared with that of bigger kernel size, which can make the training procedure more easily to be convergent.

Neural networks actually perform nonlinear operation for data. With multiple layers of processing, the neural networks can build a high-order nonlinear model for real images, which is a much more suitable representation for the data than the traditional man-crafted features. As a result, different numbers of layers also influence the property of features. In the classical ConvNet, there is only one path for the signal to go, and the classification only performs on the features extracted by the last layer of the network. In fact, features from different levels of the networks all contain useful information. Thus we propose to build a multi-stream local feature hierarchy network (LFHN). Within each stream, features of different orders are extracted by using different numbers of convolutional layers. Then features from different streams are contact to compose a multi-hierarchy feature with the size of $h \times w \times (c_1 + c_2 + \cdots + c_n)$, where $h \times w \times c_i$ is the size of the feature from stream i and c_i is the depth dimension.

In order to keep the spatial information for the local features, the convolutional kernel of 1×1 is applied. As shown in Fig. 18, there are two steams for the proposed network. One stream contains two convolutional layers where the kernel size of Conv2 and Conv3 is $1 \times 1 \times 200$ and $1 \times 1 \times 400$, respectively. In another stream, there is only one convolutional layer Conv4 with the kernel of $1 \times 1 \times 300$. Through these two streams, we can get local features from different levels of hierarchy. In order to achieve the final recognition task, one more convolution layer, Conv5 with kernel size $1 \times 1 \times 500$, and two fully connected layer are employed for the further feature abstraction.

4.3 Training

Since the root layers of convolutional networks always contain more generic features such as edges or color blob, which is useful for many tasks including face recognition, in training, we keep the pretrained results of AlexNet as the weight for the convolutional layer, Conv1. Then for the other layers, they are trained according to the Softmax loss function based on the identity labels for images in MultiPIE dataset [16].

5 Experiments

5.1 Dataset

To evaluate the effectiveness of the proposed local feature hierarchy networks (LFHN) under different poses and illumination, the MultiPIE face database [16] is employed. The MultiPIE face database contains 754,204 images of 337 identities. Each identity has images captured under 15 different poses and 20 different lighting conditions. For the original images in MultiPIE, we have aligned all the images according to the position of eyes and crop them to the size of 256×256. For each subject, we only select the images with neural expression but in all poses and lighting conditions; thus for each person, there are 300 images. Altogether, for all the individuals in the dataset, we put $300 \times 337 = 101{,}100$ images into the data pool for training and testing. In MultiPIE, there are four sessions to take photos for each subject, but not everyone comes in each session. Therefore we take all the images of 250 individuals in session 1 and the images of the other 87 persons in session 2.

5.2 Recognition Across Poses and Illumination

In order to evaluate the performance of the proposed local hierarchy networks, MultiPIE is used to train and test the networks. For the proposed networks, there is only one input instead of multiple images from different reviews. Therefore training images are randomly selected from images of natural expression but in 15 different poses and 20 different lighting conditions. For all the 337 individuals, 90,000 images are randomly taken from 101,100 images for training, and the leftover images are used as testing images. The proposed network can learn local nonlinear features which can represent the correlations between images in different poses and lighting conditions. The rank-1 recognition rates for images with pose and illumination variations are shown in Table 1. The recognition results for each view are the average results for all the images under 20 different lighting conditions.

From the results, we can see that the proposed LFHN network achieved relatively stable performance for different poses. Especially for profile-wised images, where the yaw angle is in the range of [−90°, −60°] and [60°, 90°], the average recognition rate is 97.78%, while for the traditional methods, the performance declined significantly for images with greater pose variations. For the patch-based partial recognition (PBPR) [14], the average recognition rate for front-wised images is 98.96% where the yaw angle is within 45°. But for the profile-wised images, the recognition rate is 78.76%. That indicates the projection recovery method used in PBPR does not find the accurate locations for profile-wised images. Compared with current state-of-the-art methods, the proposed LFHN network improves the recognition rate by 7.55% for images under arbitrary poses and illumination. In the proposed networks, we consider images from different views and illumination

Table 1 Rank-1 identification rates on combined variations of pose and illumination on MultiPIE

PoseID	Yaw	Pitch	RR [32]	FIP [73]	PBPR [14]	LFHN (Ours)
081	−45°	25°	24	–	88	**94.51**
110	−90°	0°	20.5	–	51	**97.52**
120	−75°	0°	26.5	–	79	**98.15**
090	−60°	0°	50.64	–	90.86	**98.51**
080	−45°	0°	65.30	67.10	**97.91**	97.75
130	−30°	0°	70.97	74.60	**99.41**	98.08
140	−15°	0°	81.07	86.10	**99.05**	97.12
050	15°	0°	77.21	83.30	**99.94**	93.74
041	30°	0°	73.69	75.30	**99.23**	97.91
190	45°	0°	58.12	61.80	**98.21**	96.53
200	60°	0°	45.97	–	87.75	**97.65**
010	75°	0°	31	–	89	**97.57**
240	90°	0°	18	–	75	**97.3**
191	45°	25°	40	–	96	**93.73**
		Mean	48.78	–	89.31	**96.86**

Table 2 Rank-1 identification rates for pose variation on MultiPIE

PoseID	Pose	PLS [54]	MCCA [48]	PLS + LDA [27]	MCCA + LDA [27]	MvDA [28]	GMA [52]	MvDN [27]	LFHN (ours)
110	−90°	0.319	0.409	0.38	0.488	0.568	0.526	0.704	**1**
120	−75°	0.775	0.742	0.798	0.662	0.723	0.723	0.822	**0.9767**
090	−60°	0.892	0.822	0.869	0.817	0.845	0.845	0.883	**1**
080	−45°	0.934	0.723	0.944	0.887	0.92	0.901	0.911	**1**
130	−30°	0.883	0.685	0.92	1	0.967	1	**0.991**	0.893
140	−15°	0.981	0.92	0.995	1	1	1	**1**	1
050	15°	0.981	0.906	0.986	1	1	1	**1**	0.938
041	30°	0.934	0.798	0.967	0.995	0.991	1	**0.991**	0.971
190	45°	0.906	0.747	0.883	0.831	0.897	0.906	0.93	**0.958**
200	60°	0.873	0.779	0.85	0.803	0.864	0.859	0.911	**0.935**
010	75°	0.723	0.714	0.709	0.676	0.714	0.718	0.798	**1**
240	90°	0.268	0.376	0.319	0.568	0.559	0.573	0.709	**1**
	Mean	0.789	0.718	0.802	0.811	0.837	0.838	0.887	**0.973**

equally. Thus pose-invariant and illumination-invariant nonlinear local features can be sought by the proposed network LFHN.

Besides the face recognition with the combined variations of pose and illumination, we also test the performance of the proposed network LFHN on pose only. For this task, the probe dataset includes images of all subjects from four sessions where images are taken in ambient lighting and 13 different poses. The comparison results with other methods are shown in Table 2.

From the results we can see even though different methods try to tackle the pose problem, the face recognition rate decreases along with the amount of pose variation. The more the view diverges from the frontal face, the lower the recognition rate is. That is because pose changes the appearance and structure of the image. MvDN [27] and MvDA [28] tried to find the correlation between different poses and achieved relatively good performance. For the proposed network LFHN, we also focus on the extraction of local features among different poses which can describe the correlation of different poses and also discriminate different identity. Thus we achieve better results compared with other methods. Especially for larger pose diversity, the performance of the proposed network is not degenerated but very stable instead. The average recognition rate is 97.3% which improves the state-of-the-art method by 8.6%.

6 Conclusion

Pose and illumination will always bring great variance for the appearance of face images, which makes face recognition across pose and illumination challenged. However, it is quite normal to encounter the pose and illumination changes in uncontrolled environment. Therefore a robust face recognition system has to deal with illumination and pose variations effectively. In fact, there are tight correlations for images from different postures. Images of different views are the projection of the same object to different positions. Then local features are more useful for recognition under different views than the global features, where the global structure is actually destructed by the projection in different views. Thus we propose a neural network which extracts local features by 1×1 convolutional kernels; in addition multi-hierarchical features are combined for the task of recognition. Experiments on MultiPIE dataset show very good and stable performance for the proposed networks in a wide range of pose and illumination.

Acknowledgements This chapter is partly supported by the National Natural Science Foundation of China (No.61502388), Ph.D. Programs Foundation of Ministry of Education of China (No. 20136102120041), the Fundamental Research Funds for the Central Universities (No. 3102015BJ (II)ZS016), and the Shaanxi Province International Science and Technology Cooperation and Exchange Program (2017KW002).

References

1. Ahonen, T., Hadid, A., Pietikainen, M.: Face description with local binary patterns: Application to face recognition. IEEE Transactions on Pattern Analysis and Machine Intelligence **28**(12), 2037–2041 (2006). DOI 10.1109/TPAMI.2006.244
2. Ahonen, T., Hadid, A., Pietikainen, M.: Face description with local binary patterns: Application to face recognition. IEEE transactions on pattern analysis and machine intelligence **28**(12), 2037–2041 (2006)
3. Andrew, G., Arora, R., Bilmes, J., Livescu, K.: Deep canonical correlation analysis. In: International Conference on Machine Learning, pp. 1247–1255 (2013)
4. Ashraf, A.B., Lucey, S., Chen, T.: Learning patch correspondences for improved viewpoint invariant face recognition. In: Computer Vision and Pattern Recognition, 2008. CVPR 2008. IEEE Conference on, pp. 1–8. IEEE (2008)
5. Basri, R., Jacobs, D.W.: Lambertian reflectance and linear subspaces. IEEE Transactions on Pattern Analysis and Machine Intelligence **25**(2), 218–233 (2003)
6. Bay, H., Ess, A., Tuytelaars, T., Van Gool, L.: Speeded-up robust features (surf). Computer vision and image understanding **110**(3), 346–359 (2008)
7. Belhumeur, P.N., Kriegman, D.J.: What is the set of images of an object under all possible illumination conditions? International Journal of Computer Vision **28**(3), 245–260 (1998)
8. Biswas, S., Aggarwal, G., Flynn, P.J., Bowyer, K.W.: Pose-robust recognition of low-resolution face images. IEEE transactions on pattern analysis and machine intelligence **35**(12), 3037–3049 (2013)
9. Castillo, C.D., Jacobs, D.W.: Using stereo matching with general epipolar geometry for 2d face recognition across pose. IEEE Transactions on Pattern Analysis and Machine Intelligence **31**(12), 2298–2304 (2009)
10. Chen, D., Cao, X., Wen, F., Sun, J.: Blessing of dimensionality: High-dimensional feature and its efficient compression for face verification. In: Proceedings of the IEEE Conference on Computer Vision and Pattern Recognition, pp. 3025–3032 (2013)
11. Ding, C., Choi, J., Tao, D., Davis, L.S.: Multi-directional multi-level dual-cross patterns for robust face recognition. IEEE Transactions on Pattern Analysis and Machine Intelligence **38**(3), 518–531 (2016). DOI 10.1109/TPAMI.2015.2462338
12. Ding, C., Tao, D.: Pose-invariant face recognition with homography-based normalization. Pattern Recognition **66**, 144–152 (2017)

13. Ding, C., Xu, C., Tao, D.: Multi-task pose-invariant face recognition. IEEE Transactions on Image Processing **24**(3), 980–993 (2015)
14. Ding, C., Xu, C., Tao, D.: Multi-task pose-invariant face recognition. IEEE Transactions on Image Processing **24**(3), 980–93 (2015)
15. Gao, Y., Leung, M.K.: Face recognition using line edge map. IEEE transactions on pattern analysis and machine intelligence **24**(6), 764–779 (2002)
16. Gross, R., Matthews, I., Cohn, J., Kanade, T., Baker, S.: Multi-pie. Image and Vision Computing **28**(5), 807–813 (2010)
17. Gunther, M., Costa-Pazo, A., Ding, C., Boutellaa, E.: The 2013 face recognition evaluation in mobile environment. 2013 International Conference on Biometrics (ICB) pp. 1–7 (2013)
18. He, X., Yan, S., Hu, Y., Niyogi, P., Zhang, H.J.: Face recognition using laplacian faces. IEEE transactions on pattern analysis and machine intelligence **27**(3), 328–340 (2005)
19. Ho, H.T., Chellappa, R.: Pose-invariant face recognition using markov random fields. IEEE transactions on image processing **22**(4), 1573–1584 (2013)
20. Hu, J., Lu, J., Tan, Y.P.: Discriminative deep metric learning for face verification in the wild. In: Proceedings of the IEEE Conference on Computer Vision and Pattern Recognition, pp. 1875–1882 (2014)
21. Huang, S.C., Cheng, F.C., Chiu, Y.S.: Efficient contrast enhancement using adaptive gamma correction with weighting distribution. IEEE Transactions on Image Processing **22**(3), 1032–1041 (2013)
22. Jiang, X., Cheng, Y., Xiao, R., Li, Y., Zhao, R.: Spherical harmonic based linear face de-lighting and compensation. Applied Mathematics and Computation **185**(2), 857–868 (2007). https://doi.org/10.1016/j.amc.2006.06.090. http://www.sciencedirect.com/science/article/pii/S0096300306007673. Special Issue on Intelligent Computing Theory and Methodology
23. Jiang, X., Feng, X., Wu, J., Peng, J.: Lighting alignment for image sequences. In: International Conference on Image and Graphics, pp. 462–474. Springer (2015)
24. Jiang, X., Kong, Y.O., Huang, J., Zhao, R., Zhang, Y.: Learning from real images to model lighting variations for face images. In: European Conference on Computer Vision (ECCV), pp. 284–297 (2008)
25. Kafai, M., An, L., Bhanu, B.: Reference face graph for face recognition. IEEE Transactions on Information Forensics and Security **9**(12), 2132–2143 (2014)
26. Kan, M., Shan, S., Chang, H., Chen, X.: Stacked progressive auto-encoders (spae) for face recognition across poses. In: Proceedings of the IEEE Conference on Computer Vision and Pattern Recognition, pp. 1883–1890 (2014)
27. Kan, M., Shan, S., Xilin., C.: Multi-view deep network for cross-view classification. In: IEEE Conference on Computer Vision and Pattern Recognition (2016)
28. Kan, M., Shan, S., Zhang, H., Lao, S., Chen, X.: Multi-view discriminant analysis. IEEE transactions on pattern analysis and machine intelligence **38**(1), 188–194 (2016)
29. Krizhevsky, A., Sutskever, I., Hinton, G.E.: Imagenet classification with deep convolutional neural networks. In: Advances in neural information processing systems, pp. 1097–1105 (2012)
30. Lee, K.C., Ho, J., Kriegman, D.: Acquiring linear subspaces for face recognition under variable lighting. IEEE Transactions on Pattern Analysis and Machine Intelligence **27**(5), 684–698 (2005)
31. Li, A., Shan, S., Chen, X., Gao, W.: Maximizing intra-individual correlations for face recognition across pose differences. In: Computer Vision and Pattern Recognition, 2009. CVPR 2009. IEEE Conference on, pp. 605–611. IEEE (2009)
32. Li, A., Shan, S., Gao, W.: Coupled bias-variance tradeoff for cross-pose face recognition. IEEE Transactions on Image Processing **21**(1), 305–15 (2012)
33. Li, H., Hua, G., Lin, Z., Brandt, J., Yang, J.: Probabilistic elastic matching for pose variant face verification. In: Proceedings of the IEEE Conference on Computer Vision and Pattern Recognition, pp. 3499–3506 (2013)
34. Li, S., Liu, X., Chai, X., Zhang, H., Lao, S., Shan, S.: Morphable displacement field based image matching for face recognition across pose. European Conference on Computer Vision 2012 pp. 102–115 (2012)

35. Liao, Q., Leibo, J.Z., Poggio, T.: Learning invariant representations and applications to face verification. In: Advances in Neural Information Processing Systems, pp. 3057–3065 (2013)
36. Liao, S., Jain, A.K., Li, S.Z.: Partial face recognition: Alignment-free approach. IEEE Transactions on pattern analysis and machine intelligence **35**(5), 1193–1205 (2013)
37. Lowe, D.G.: Distinctive image features from scale-invariant keypoints. International journal of computer vision **60**(2), 91–110 (2004)
38. Majumdar, A., Singh, R., Vatsa, M.: Face recognition via class sparsity based supervised encoding. IEEE transactions on pattern analysis and machine intelligence (2016)
39. Masi, I., Rawls, S., Medioni, G., Natarajan, P.: Pose-aware face recognition in the wild. In: Proceedings of the IEEE Conference on Computer Vision and Pattern Recognition, pp. 4838–4846 (2016)
40. Parkhi, O.M., Vedaldi, A., Zisserman, A., et al.: Deep face recognition. In: British Machine Vision Conference, vol. 1, p. 6 (2015)
41. Peng, X., Yu, X., Sohn, K., Metaxas, D., Chandraker, M.: Reconstruction for feature disentanglement in pose-invariant face recognition. arXiv preprint arXiv:1702.03041 (2017)
42. Pentland, A., Moghaddam, B., Starner, T., et al.: View-based and modular eigenspaces for face recognition. In: CVPR, vol. 94, pp. 84–91 (1994)
43. Pizer, S.M., Amburn, E.P., Austin, J.D., Cromartie, R., Geselowitz, A., Greer, T., ter Haar Romeny, B., Zimmerman, J.B., Zuiderveld, K.: Adaptive histogram equalization and its variations. Computer vision, graphics, and image processing **39**(3), 355–368 (1987)
44. Prince, S.J., Elder, J.H., Warrell, J., Felisberti, F.M.: Tied factor analysis for face recognition across large pose differences. IEEE Transactions on pattern analysis and machine intelligence **30**(6), 970–984 (2008)
45. Ramamoorthi, R., Hanrahan, P.: On the relationship between radiance and irradiance: determining the illumination from images of a convex lambertian object. Journal of the Optical Society of America, A **18**(10), 2448–2459 (2001)
46. Ren, S., He, K., Girshick, R., Sun, J.: Faster r-cnn: Towards real-time object detection with region proposal networks. Advances in Neural Information Processing Systems (NIPS) pp. 91–99 (2015)
47. Rupnik, J., Shawe-Taylor, J.: Multi-view canonical correlation analysis. In: Conference on Data Mining and Data Warehouses (SiKDD 2010), pp. 1–4 (2010)
48. Rupnik, J., Shawe-Taylor, J.: Multi-view canonical correlation analysis. Conference on Data Mining and Data Warehouses(SiKDD 2010) pp. 1–4 (2010)
49. Schölkopf, B., Smola, A., Müller, K.R.: Nonlinear component analysis as a kernel eigenvalue problem. Neural computation **10**(5), 1299–1319 (1998)
50. Schroff, F., Kalenichenko, D., Philbin, J.: Facenet: A unified embedding for face recognition and clustering. Computer Vision and Pattern Recognition pp. 815–823 (2015)
51. Schroff, F., Treibitz, T., Kriegman, D., Belongie, S.: Pose, illumination and expression invariant pairwise face-similarity measure via doppelgänger list comparison. In: Computer Vision (ICCV), 2011 IEEE International Conference on, pp. 2494–2501. IEEE (2011)
52. Sharma, A.: Generalized multiview analysis: A discriminative latent space. In: IEEE Conference on Computer Vision and Pattern Recognition, pp. 2160–2167 (2012)
53. Sharma, A., Al Haj, M., Choi, J., Davis, L.S., Jacobs, D.W.: Robust pose invariant face recognition using coupled latent space discriminant analysis. Computer Vision and Image Understanding **116**(11), 1095–1110 (2012)
54. Sharma, A., Jacobs, D.W.: Bypassing synthesis: Pls for face recognition with pose, low-resolution and sketch. In: IEEE Conference on Computer Vision and Pattern Recognition, pp. 593–600 (2011)
55. Sharma, A., Kumar, A., Daume, H., Jacobs, D.W.: Generalized multiview analysis: A discriminative latent space. In: Computer Vision and Pattern Recognition (CVPR), 2012 IEEE Conference on, pp. 2160–2167. IEEE (2012)
56. Shashua, A., Riklin-Raviv, T.: The quotient image: Class-based re-rendering and recognition with varying illuminations. IEEE Transactions on Pattern Analysis and Machine Intelligence **23**(2), 129–139 (2001)

57. Sun, Y., Wang, X., Tang, X.: Deeply learned face representations are sparse, selective, and robust. Computer Vision and Pattern Recognition
58. Szegedy, C., Liu, W., Jia, Y., Sermanet, P., Reed, S., Anguelov, D., Erhan, D., Vanhoucke, V., Rabinovich, A.: Going deeper with convolutions. In: The IEEE Conference on Computer Vision and Pattern Recognition (CVPR) (2015)
59. Taigman, Y., Yang, M., Ranzato, M., Wolf, L.: Deepface: Closing the gap to human-level performance in face verification. In: Proceedings of the IEEE Conference on Computer Vision and Pattern Recognition, pp. 1701–1708 (2014)
60. Tan, X., Triggs, B.: Enhanced local texture feature sets for face recognition under difficult lighting conditions. IEEE transactions on image processing **19**(6), 1635–1650 (2010)
61. Tran, L., Yin, X., Liu, X.: Disentangled representation learning gan for pose-invariant face recognition. In: IEEE Conference on Computer Vision and Pattern Recognition, vol. 3, p. 7 (2017)
62. Turk, M., Pentland, A.: Eigenfaces for recognition. Journal of cognitive neuroscience **3**(1), 71–86 (1991)
63. Wagner, A., Wright, J., Ganesh, A., Zhou, Z., Mobahi, H., Ma, Y.: Toward a practical face recognition system: robust alignment and illumination by sparse representation. IEEE Transactions on Pattern Analysis and Machine Intelligence **34**(2), 372–386 (2012)
64. Wang, H., Li, S.Z., Wang, Y.: Generalized quotient image. In: Computer Vision and Pattern Recognition, 2004. CVPR 2004. Proceedings of the 2004 IEEE Computer Society Conference on, vol. 2, pp. II–II. IEEE (2004)
65. Wang, X., Han, T.X., Yan, S.: An hog-lbp human detector with partial occlusion handling. In: 2009 IEEE 12th International Conference on Computer Vision, pp. 32–39 (2009). DOI 10.1109/ICCV.2009.5459207
66. Wen, Y., Zhang, K., Li, Z., Qiao, Y.: A discriminative feature learning approach for deep face recognition. In: B. Leibe, J. Matas, N. Sebe, M. Welling (eds.) Computer Vision – ECCV 2016, pp. 499–515. Springer International Publishing, Cham (2016)
67. Wiskott, L., Krüger, N., Kuiger, N., Von Der Malsburg, C.: Face recognition by elastic bunch graph matching. IEEE Transactions on pattern analysis and machine intelligence **19**(7), 775–779 (1997)
68. Xie, X., Zheng, W.S., Lai, J., Yuen, P.C., Suen, C.Y.: Normalization of face illumination based on large-and small-scale features. IEEE Transactions on Image Processing **20**(7), 1807–1821 (2011)
69. Yang, J., Frangi, A.F., Yang, J.y., Zhang, D., Jin, Z.: Kpca plus lda: a complete kernel fisher discriminant framework for feature extraction and recognition. IEEE Transactions on pattern analysis and machine intelligence **27**(2), 230–244 (2005)
70. Zhang, Y., Shao, M., Wong, E.K., Fu, Y.: Random faces guided sparse many-to-one encoder for pose-invariant face recognition. In: Proceedings of the IEEE International Conference on Computer Vision, pp. 2416–2423 (2013)
71. Zhao, W., Chellappa, R., Phillips, P.J., Rosenfeld, A.: Face recognition: A literature survey. ACM Computing Surveys **35**(4), 399–458 (2003)
72. Zhou, H., Sadka, A.H.: Combining perceptual features with diffusion distance for face recognition. IEEE Transactions on Systems, Man, and Cybernetics, Part C (Applications and Reviews) **41**(5), 577–588 (2011)
73. Zhu, Z., Luo, P., Wang, X., Tang, X.: Deep learning identity-preserving face space. In: Proceedings of the IEEE International Conference on Computer Vision, pp. 113–120 (2013)
74. Zhu, Z., Luo, P., Wang, X., Tang, X.: Multi-view perceptron: a deep model for learning face identity and view representations. Advances in Neural Information Processing Systems (NIPS) pp. 217–225 (2014)

Face Anti-spoofing via Deep Local Binary Pattern

Lei Li and Xiaoyi Feng

Abstract In recent years, convolutional neural network (CNN) has achieved satisfactory performance in computer vision and pattern recognition. When we visualize the convolutional responses, we can conclude that the convolutional responses include some diacritically structural information. But for the high dimensionality of them, it is not feasible to directly use the responses to detect fake faces. Moreover, the small size of existing face anti-spoofing databases leads to the difficulty of training a new CNN model. Compared with deep learning, the traditional handcrafted features, such as local binary pattern (LBP), have been successfully used in face anti-spoofing and achieved good detection results. So in our work, we extracted the handcrafted features from the convolutional responses of the fine-tuned CNN model. More specifically, the CNN is first fine-tuned based on a pre-trained VGG-face model. Then, the LBP features are calculated from the convolutional responses and concatenated into one feature vector. After that, the final vectors are fed into a support vector machine (SVM) classifier to detect the fake faces. Validated on two public available databases, Replay-Attack and CASIA-FA, our proposed detection method can obtain promising results compared to the state-of-the-art methods.

1 Introduction

At present, a variety of face spoofing attacks have become a major security threat to most face recognition systems. To make matters worse, with the development of the Internet and social media, criminals can easily learn how to invade a biometric system. For instance, a fingerprint hack in YouTube performed how to unlock the iPhone5S Touch ID preconization system. A face spoofing attack occurs when someone tries to bypass a face biometric system by presenting a fake face in front

---

L. Li · X. Feng (✉)
School of Electronics and Information, Northwestern Polytechnical University, Xi'an, Shaanxi, China
e-mail: lilei_npu@mail.nwpu.edu.cn; fengxiao@nwpu.edu.cn

X. Jiang et al. (eds.), *Deep Learning in Object Detection and Recognition*,
https://doi.org/10.1007/978-981-10-5152-4_4

of the camera. For instance, researchers inspected the threat of the online social networks based on facial disclosure against the latest version of six commercial face authentication systems (Face Unlock, Facelock Pro, Visidon, VeriFace, Luxand Blink, and FastAccess). While on average only 39% of the images published on social networks can be successfully used for spoofing, the relatively small number of usable images was enough to fool face authentication software of 77% of the 74 users [1]. Also, in a live demonstration during the International Conference on Biometric (ICB 2013), a female intruder with a specific makeup succeeded in fooling a face recognition system. These examples among many others highlight the vulnerability of face recognition systems to spoofing attacks. Based on different materials of fake faces, three kinds of face spoofing attacks can be considered: (i) print attacks, (ii) replay attacks, and (iii) 3D mask attacks. Print attacks use still photos or images of legitimate users to spoof face biometric systems. Replay attacks utilize face videos to intrude the systems. These two kinds of attacks belong to 2D face attacks. However, 3D mask attacks are often launched by a 3D face mask, which needs complicated production process and high costs. In these attacks, the 2D attacks are the most common and simplest face spoofing attacks. According to the frequency of attacks and the situation of practical application statistics, in this chapter, we focus on print attacks and replay attacks by the extended detection method [2].

To detect face spoofing attacks, many methods have been proposed in the recent decade. Based on different learning algorithms, they can be classified into machine learning-based methods and deep learning-based methods. For the former, they mainly extract the handcrafted features from face images and train a classifier to distinguish the real and fake faces. Although they can get satisfactory performance in some specific cases, they need the researchers to design the features with prior knowledge. Moreover, the handcrafted features cannot promise us the methods can also get good performance in unknown environment. Compared with machine learning-based methods, deep learning-based methods focus on designing the structure of the model and are more robust [3] with its various frequency responses. Generally speaking, a more deep model can get better performances than the shallow one. But it requires many training samples to optimize the parameters, which is unpractical for the existing face anti-spoofing databases.

As aforementioned, the features extracted by deep learning are robust mainly caused by the various frequency responses, and the handcrafted features have more powerful ability of characterizing the image. When we extract the handcrafted features from the various frequency responses of convolutional layers, the obtained features are more robust. Therefore, we combine machine learning and deep learning together in our proposed method. Moreover, the real and fake faces are very distinctive in chrominance channels [4]. So, we also exploit the color information to detect face spoofing. More specifically, we first fine-tune a pre-trained deep learning model with the existing face anti-spoofing databases and then extract the color LBP features from the convolutional responses. At the end, we concatenate the LBP features into a feature vector and train an SVM classifier to detect the fake faces.

We validated our work on Replay-Attack and CASIA-FA databases by intra test and cross test. Experiments show that the performance of our method is comparable to the state-of-the-art methods.

2 Related Work

In this section, we review typical nonintrusive software-based countermeasure for face anti-spoofing. Based on different cues, prior detection approaches can be classified into (1) texture analysis based, (2) motion analysis based, (3) image quality analysis based, and (4) multi-cues fused based.

Texture Analysis-Based Methods Usually, the fake face is invoked to printed photos, mobile phones, and other electronic equipments. However, there is a significant difference between these mediums and the real faces. For example, these mediums are 2D planes, while the real faces are 3D planes, which will lead to the discrepancy of light's diffuse reflection. So, in [5], the specular reflection features were extracted to distinguish the 2D spoofing attacks and the real faces. Moreover, for the 2D attacks, the fake faces are printed by the printers or showed by the LCD screens. But these devices are still imperfect in color gamut compared with real world, which will lead to the subtle differences in texture. So, Maatta et al. [6] concatenated various LBP features to detect the fake faces. Unlike [6], Chingovska et al. [7] extracted the LBP features from different blocks of a gray-scaled image. First, they divided the face image into several blocks and extracted the LBP features from each block. To capture the structural information, those LBP features were concatenated into one feature histogram, which will be fed into a classifier to distinguish the real and fake faces. In another work [4], the LBP features were extracted from different color spaces (RGB, HSV, and YCbCr) and were fed into an SVM classifier to detect the spoofing attacks. After that, they [8] extended the texture features from LBP to other descriptors, such as co-occurrence of adjacent local binary patterns (CoALBP), local phase quantization (LPQ), and binarized statistical image features (BSIF). Deep learning is a hot topic for its superior performance than the traditional machine learning. Under this trend, Yang et al. in [9], and Menotti et al. in [10], explored the effectiveness of deep learning on face anti-spoofing. The former designed a CNN model based on an existing network structure [11], while the latter designed the network structure by themselves. In these works, they train their CNN models using the existing face anti-spoofing databases. However, the scale of these anti-spoofing databases is quite small, and they are collected in controlled environments. Thus, it is quite hard to train a deep model. Although the texture analysis based methods are able to get satisfactory performance, there are still some limitations. For example, the input face images are required to have high resolution, which means the acquisition systems must have good quality.

Motion Analysis-Based Methods Based on the fact that print photo or image attacks have no living information (such as blinking), some researchers use different methods to capture the motion information. For instance, Anjos et al. [12] analyzed the relative movement between the face region and the background to distinguish the real and fake faces. Because the real face regions always have some motion compared with the background, the fake face regions are relative still. In another, Pereira et al. [13] calculated the local binary patterns from three orthogonal planes (LBP-TOP) to represent dynamic textures. Different from traditional LBP, the LBP-TOP features combine texture information and motion information. It is noted that there is no operator of face alignment, which means the motion information not only includes the facial key point motion but also the head motion. At the end, the LBP-TOP features are fed into a classifier to detect face spoofing. In addition, Kollreider et al. [14] presented a new method based on simplified optical flow analysis. They selected some key facial parts (e.g., eyes/nose, left and right ear) and computed the changes of their optical flow. Then, a binary detector was used to evaluate the trajectories of selected parts. Also, Chakka et al. [15] used the optical flow algorithm to detect these mediums (such as paper or screen). The main problems of motion-based methods are that the verification process takes some time and the real client needs to be very cooperative in practice. Even though motion is an important visual cue, the lack of motion may lead to a high number of authentication failures if user cooperation is not requested.

Image Quality Analysis-Based Methods When launching a face attack, the fake face is prepared by passing through two different camera systems and a printing system or display device. Thus, the fake face actually can be regarded as a recaptured image. However, due to the imperfection of these devices, the recaptured images always present low image quality compared with the real face image. Therefore, Zhang et al. [16] considered the high-frequency information was the response of image quality. So they extracted the image quality features to detect the fake faces by multiple different Gaussian filters. In another work, Peng et al. [17] also detected fake face by extracting the high-frequency information. But different with [16], they added a flash in front of the camera. By calculating the changes of high-frequency information with and without the flash, it can tackle the problem of face anti-spoofing. If the changes are greater than a pre-set threshold, the face image will be regarded as a fake face. In another work, Wen et al. [5] extracted four kinds of features (such as specular reflection features, blurriness features, chromatic moment features, and color diversity features) to describe the image quality. Then these features were concatenated into a feature vector and fed into an SVM classifier to detect face spoofing attacks. Also, Galbally et al. [18] measured the image quality by 25 kinds of assessment criteria, including 21 full-reference measures and 4 non-reference measures, to detect the fake faces. The image quality analysis-based methods can work well if the fake faces have lower image quality. But with the improvement of the printing equipments and display devices, the image quality of fake faces usually has subtle differences compared with real faces.

Multi-cues Fuse-Based Methods Compared to face texture or motion, some other biological information (such as skin, voice, fingerprint, and iris) can also be applied to distinguish the real and fake clients. In [19], Zhang et al. proposed to analyze the reflectance property by Lambertian reflectance model. Girija [20] fused the face and voice features to check liveness. Wang et al. [21] proposed a new face anti-spoofing method by recovering sparse 3D facial structure. Recently, Akhtar et al. [22] concatenated face, iris, and fingerprint features to detect spoofing attacks. In another work [23], they implemented a mobile biometric liveness detection system. Base on the different level of security, the system can select disparate anti-methods to meet the demand. However, these methods need extra requirements on the user or the biometric system. As a consequence, the application range is very narrow.

3 Proposed Method

In our work, we propose to extract the color LBP descriptors from the convolutional feature maps. The main architecture of the proposed method is illustrated in Fig. 1.

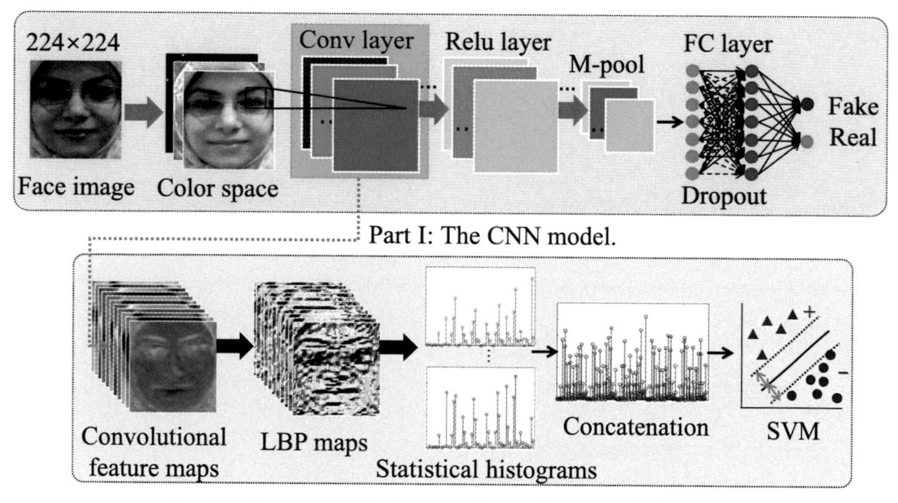

Fig. 1 Architecture of face anti-spoofing based on deep local binary pattern. Part I is the main structure of the fine-tuned CNN model. Part II is the main idea of extracting the LBP features from the convolutional feature maps and concatenating them into one feature vector. In Part I, Conv Layer is convolutional layer, Relu Layer is rectified linear units layer, M-pool is pooling layer with max value operation, and FC Layer is the fully connected layer

3.1 CNN Training

Unlike other deep learning works [9, 24], the existing pre-trained deep learning model called VGG-face is fine-tuned to reduce the influence of over-fitting. It is noted that the VGG-face model is designed for face recognition [25], and its network parameters are given in Table 1. For VGG-face model, it is trained with a large set of face images (2.6M images) and tested on two challenging face recognition databases: Labeled Face in the World and Youtube face achieving state-of-the-art results. Considering VGG-face is used to recognize 2622 subjects, we should change the output of the last fully connected layer from 2622 to 2. Here 2 is the classes of face: real and fake. The softmaxloss function illustrated in Eq. 1 is used as the cost function when fine-tuning the parameters of VGG-face model.

$$\mathscr{F}(Y) = \sum_{i=1}^{n} \{\log(e^{y_{i1}} + e^{y_{i2}} + \ldots + e^{y_{iv}}) + y_{ir}\} \tag{1}$$

where i is the index of training samples and n is the number of training samples. $Y = [Y_1, Y_2, \ldots, Y_i, \ldots, Y_n]$ is the set of labels, $Y_i = [y_{i1}, y_{i2}, \ldots, y_{ir}, \ldots, y_{ik}]$ is the predict vector of the ith sample, and v is the number of classes. It is noted that the y_{ir} is the predict value of the ith sample.

3.2 Color Spaces

RGB is the most used color space for sensing, representation, and displaying of color images. Even though it includes three color channels (red, green, and blue), RGB does not consider the information of luminance and chrominance. Therefore, we feed the VGG-face model with the face images in two other color spaces: the HSV and the YCbCr. Different from RGB color space, the HSV and YCbCr color spaces are based on the separation of the luminance and the chrominance information. More specifically, in the HSV color space, the hue and the saturation dimensions define the chrominance of the image, while the value dimension corresponds to the luminance. The YCbCr space separates the RGB components into luminance (Y), chrominance blue (Cb), and chrominance red (Cr). More details about these color spaces have been illustrated in [26].

3.3 Local Binary Pattern (LBP)

For face analysis, LBP [27] is a very effective texture descriptor. For each pixel in an image, a binary code is computed by thresholding a circularly symmetric neighborhood with the value of the central pixel. As aforementioned, the convo-

Table 1 The configuration parameters in the pre-trained VGG-face model. In this model, we have changed the output of the last fully connected layer from 2622 to 2 and added one dropout layer between two fully connected layers

Layer	1	2	3	4	5	6	7	8	9	10	11	12	13	14	15	16	17	18	19	20
Type	conv	relu	conv	relu	mpool	conv	relu	conv	relu	mpool	conv	relu	conv	relu	conv	relu	mpool	conv	relu	conv
Support	3	1	3	1	2	3	1	3	1	2	3	1	3	1	3	1	2	3	1	3
Filt dim	3	–	64	–	–	64	–	128	–	–	128	–	256	–	256	–	–	256	–	512
Num filts	64	–	64	–	–	128	–	128	–	–	256	–	256	–	256	–	–	512	–	512
Stride	1	1	1	1	2	1	1	1	1	2	1	1	1	1	1	1	2	1	1	1
Pad	1	0	1	0	0	1	0	1	0	0	1	0	1	0	1	0	0	1	0	1

Layer	21	22	23	24	25	26	27	28	29	30	31	32	33	34	35	36	37	38	39	40
Type	relu	conv	relu	mpool	conv	relu	conv	relu	conv	relu	mpool	conv	relu	dropout	fc	relu	dropout	fc	softmax	–
Support	1	3	1	2	3	1	3	1	3	1	2	7	1	–	1	1	–	1	1	–
Filt dim	–	512	–	–	512	–	512	–	512	–	–	512	–	–	4096	–	–	4096	–	–
Num filts	–	512	–	–	512	–	512	–	512	–	–	4096	–	–	4096	–	–	2	–	–
Stride	1	1	1	2	1	1	1	1	1	1	2	1	1	–	1	1	–	1	1	–
Pad	1	0	1	0	0	1	0	1	0	0	1	0	1	–	1	0	–	0	0	0

lutional feature maps can be regarded as the gray-scaled images. For the sake of simplicity, we use X_i to denote the index of training samples, and X_i^{RGB}, X_i^{HSV} and X_i^{YCbCr} denote the RGB, HSV, and YCbCr color images of X_i, respectively. For face image X_i, the jth convolutional layer's feature maps can been illustrated as $C_{X_i}^j = [C_{X_i}^{j1}, C_{X_i}^{j2}, \ldots, C_{X_i}^{jd}, \ldots, C_{X_i}^{jD}]$, where D is the number of the filters in the jth convolutional layer of VGG-face model.

For each pixel (x, y) in $C_{X_i}^{jd}$, the LBP code is computed as shown in Eq. 2.

$$LBP_{P,R}(x, y) = \sum_{\upsilon=1}^{P} \delta(r_\upsilon - r_c) \times 2^{n-1} \tag{2}$$

where $\delta(x) = 1$ if $x \geq 0$; otherwise $\delta(x) = 0$. r_c and $r_\upsilon(\upsilon = 1, \ldots, P)$ denote the intensity values of the central pixel (x, y) and its P neighbourhood pixels located at the circle of radius $R(R > 0)$, respectively. The occurrences of the different binary patterns are collected into histogram to represent the image texture information. LBP pattern is defined as uniform if its binary code contains at most two transitions from 0 to 1 or from 1 to 0. For example, 01110000 (2 transitions) and 00000000 (0 transitions) are uniform patterns.

To summarize, let $H_{C_{X_i}^{jd}}$ be the texture histogram extracted from $C_{X_i}^{jd}$. For the image X_i, all texture histograms are shown in Eq. 3.

$$H(X_i) = \{H_{C_{X_i}^1}, H_{C_{X_i}^2}, \ldots, H_{C_{X_i}^j}, \ldots, H_{C_{X_i}^J}\} \tag{3}$$

J is the number of convolutional layers in the fine-tuned VGG-face model. So $H(X_i^{RGB})$, $H(X_i^{HSV})$, and $H(X_i^{YCbCr})$ are the color texture histograms from X_i^{RGB}, X_i^{HSV}, and X_i^{YCbCr}, respectively.

3.4 Concatenating the LBP

Let $X = [X_1, X_2, \ldots, X_i, \ldots, X_n]$ denotes the training samples. Each convolutional feature map will get one texture histogram. In our work, we concatenate the texture histograms that belong to the same convolutional layer. So for each convolutional layer, we will obtain the concatenated LBP descriptions shown in Eq. 4.

$$\mathbb{F}^j = \begin{bmatrix} F^j_{X_1} & \dots & F^j_{X_i} & \dots & F^j_{X_n} \end{bmatrix}$$
$$= \begin{bmatrix} H_{C^{j1}_{X_1}} & \dots & H_{C^{jd}_{X_1}} & \dots & H_{C^{jD}_{X_1}} \\ H_{C^{j1}_{X_2}} & \dots & H_{C^{jd}_{X_2}} & \dots & H_{C^{jD}_{X_2}} \\ \dots & \dots & \dots & \dots & \dots \\ H_{C^{j1}_{X_i}} & \dots & H_{C^{jd}_{X_i}} & \dots & H_{C^{jD}_{X_i}} \\ \dots & \dots & \dots & \dots & \dots \\ H_{C^{j1}_{X_n}} & \dots & H_{C^{jd}_{X_n}} & \dots & H_{C^{jD}_{X_n}} \end{bmatrix} \quad (4)$$

3.5 Classification

After extracting the LBP descriptions from the convolutional feature maps, the support vector machine (SVM) is trained and invoked for face anti-spoofing. In this chapter, the LIBLINEAR toolkit [28] is used.

4 Experimental Data and Setup

4.1 Experimental Data

In this chapter, we validate our proposed method with extensive experiments on two public face anti-spoofing databases: Replay-Attack database [29] and CASIA Face Anti-spoofing database [30]. A description of these two databases is given as follows:

4.1.1 Replay-Attack

The IDIAP Replay-Attack database[1] [29] consists of 1300 video clips of real and attack and its attempts to 50 clients (Fig. 2 shows some example of real and fake faces). In the database, the clients are divided into three subject-disjoint subsets for training, development, and testing (15, 15, and 20, respectively). The genuine videos are recorded under two different lighting conditions: controlled and adverse. Two types of attacks are created: replay attacks and print attacks. In the replay attacks, the iPhone 3GS and iPad display devices are used to replay high-quality video and images of the real client. For the print attacks, the high-quality images were printed on A4 papers and presented in front of the camera.

[1] https://www.idiap.ch/dataset/replayattack/download-proc

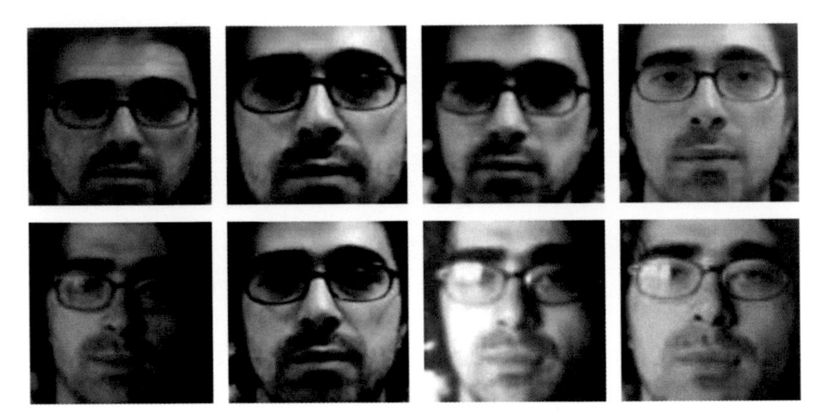

Fig. 2 Samples from the Replay-Attack database. The first row presents images taken from the controlled scenario, while the second row corresponds to the images from the adverse scenario. From left to right: real faces and the corresponding high definition, mobile, and print attacks

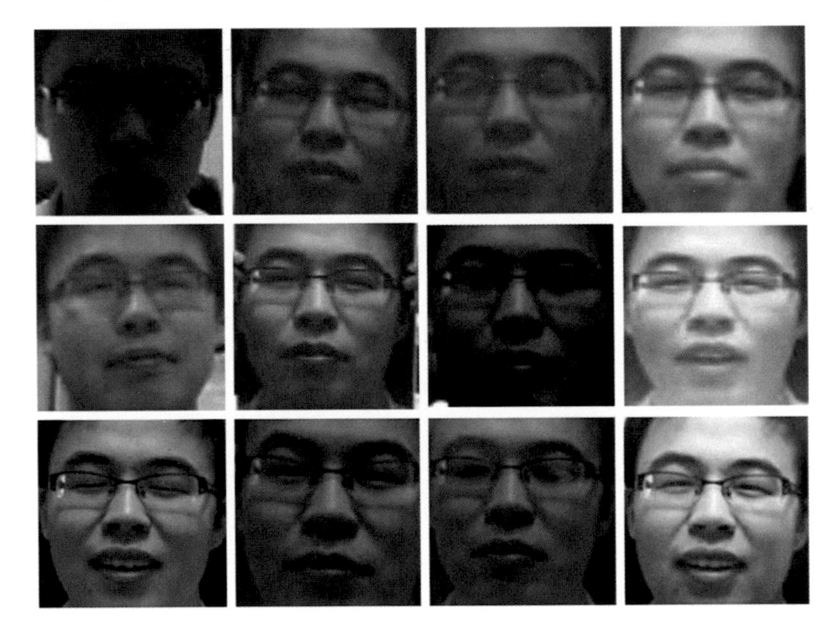

Fig. 3 Samples from the CASIA-FA. From top to bottom: low-, normal-, and high-quality images. From left to right: real faces and the corresponding warped photo, cut photo, and video replay attacks

4.1.2 CASIA-FA

The CASIA Face Anti-spoofing (CASIA-FA) database[2] [30] consists of 600 video recordings of real and attack attempts to 50 clients (Fig. 3 shows some example of

[2]http://www.cbsr.ia.ac.cn/english/FaceAntiSpoofDatabases.asp

real and fake faces), which are divided into two subject-disjoint subsets for training and testing (20 and 30, respectively). Three types of attacks are created: video replay attacks, warped attacks, and cut attacks. The real and the attack attempts were recorded using three camera devices: low, normal, and high resolution.

4.2 Experimental Setups

4.2.1 Evaluation Protocol

For the performance evaluation, we followed the overall protocol associated with the two databases. For each database, we used the training set to fine-tune the VGG-face model and the testing set to evaluate the performance. On CASIA-FA database, the results are evaluated in terms of equal error rate. The Replay-Attack database provides also a development set to tune the model parameters. Thus, the results are reported in terms of equal error rate (EER) on the development set and half total error rate (HTER) on the test set, illustrated in Eq. 5. In our experiments, we have reported two different kinds of performances: (i) the impact of different face regions for anti-spoofing and (ii) the results of cascading deep features.

$$HTER = \frac{FRR(\kappa, \mathscr{D}) + FAR(\kappa, \mathscr{D})}{2} \tag{5}$$

where $\mathscr{D}$ denotes the used database and the value of κ is estimated on the EER. $FRR(\kappa, \mathscr{D})$ means the false rejection rate of real faces, and $FAR(\kappa, \mathscr{D})$ means the false acceptance rate of fake faces.

4.2.2 Data Processing

In our proposed method, the dimension of the texture histogram is 59. For a convolutional layer, the dimension of the concatenated LBP descriptions is $59 \times D$, where D is the number of convolutional feature maps in the corresponding layer. For example, there are 64 feature maps in the first convolutional layer. So the dimension is $59 \times 64 = 3776$. It is noted that the scale of the feature maps becomes more and more small with the increase of the convolutional layer's depth. Moreover, we used the $LBP_{8,1}$ operator (i.e., $P = 8$ and $R = 1$) to extract the textural features from the convolutional feature maps. Hence, we select $j = \{1, 3, 6, 8, 11, 13, 15, 18, 20, 22\}$.

Compared to the dimensionality with the quantity of the training set, it is easy to over-fit the SVM classifier. So we use principal component analysis (PCA) algorithm to reduce the feature dimensionality. Furthermore, in order to capture both of the texture and the motion variations of the face images, the extracted features of a certain time window (two seconds) are averaged as the final feature description.

4.2.3 Intra Test and Cross Test

To evaluate the effectiveness and generalization capabilities, in our work, we carry out two kinds of test: intra test and cross test. For the former, the CNN model and the SVM classifier are trained and tested on the same database. For the latter, the CNN model and the SVM classifier are trained on one database and tested on another database, like performed in [31]. For example, we will train the CNN and the SVM on Replay-Attack database and test them on the CASIA database and vice versa.

5 Results and Discussion

In this section, we present and discuss the results obtained by our proposed method. Firstly, we feed the VGG-face model with the face images in different color spaces and extract LBP features from different convolutional layers. Apart from intra test, we also perform cross test to evaluate the generalization capability. Secondly, we concatenate the LBP of different color spaces and analysis the influence of different concatenation mechanisms. Finally, we compare the performance of proposed detection method with the state-of-the-art methods.

5.1 Impact of Different Color Spaces

5.1.1 Intra Test

In this part, we present the performance of the LBP extracted from different convolutional layers. Table 2 presents the results of the LBP descriptor applied on the different convolutional feature maps and different color spaces. For Replay-Attack database, the best of EER and HTER obtained by the RGB LBP are 0.3%

Table 2 The intra test results from different convolutional feature maps

Database			1 conv	3 conv	6 conv	8 conv	11 conv	13 conv	15 conv	18 conv	20 conv	22 conv
Replay-Attack	RGB	EER	17.8	5.2	7.9	7.2	2.8	1.5	**0.3**	2.7	2.8	4.7
		HTER	17.9	8.8	7.9	7.8	2.6	1.2	**0.9**	3.4	3.7	5.7
	HSV	EER	4.0	4.7	6.7	5.4	3.3	2.9	2.2	6.0	7.1	6.1
		HTER	3.1	5.2	8.1	5.3	2.7	3.5	3.4	7.2	7.4	8.6
	YCbCr	EER	20.5	24.9	6.7	10.9	5.4	4.2	1.6	4.6	5.4	4.8
		HTER	10.0	13.0	12.2	9.2	5.4	5.4	7.2	6.6	7.2	9.1
CASIA-FA	RGB	EER	12.1	19.9	6.4	13.9	4.3	4.3	5.1	**3.4**	4.0	5.3
	HSV	EER	10.0	13.0	12.2	9.2	5.4	5.4	7.2	6.6	7.2	9.1
	YCbCr	EER	10.2	18.0	11.1	13.0	7.2	6.1	6.4	7.3	9.0	10.0

and 0.9%, respectively, which is better than the HSV and YCbCr color spaces. Compared with the Replay-Attack database, the best of EER of CASIA-FA is 3.4%, which is also obtained in RGB color space. It can be clearly seen that the LBP extracted from the RGB color space can get the best performance compared with the HSV and YCbCr color spaces. Furthermore, we can find that the best convolutional layers which get the best EER and HTER are the 15th and 18th, rather than the deepest layer 22th. This indicates the deeper layer may not be the best layer when it extracts the handcrafted features from the convolutional feature maps.

Apart from Table 2, we plot the ROC curves of different color spaces as shown in Figs. 4 and 5. From the Fig. 4, we can find the line that possesses the shortest

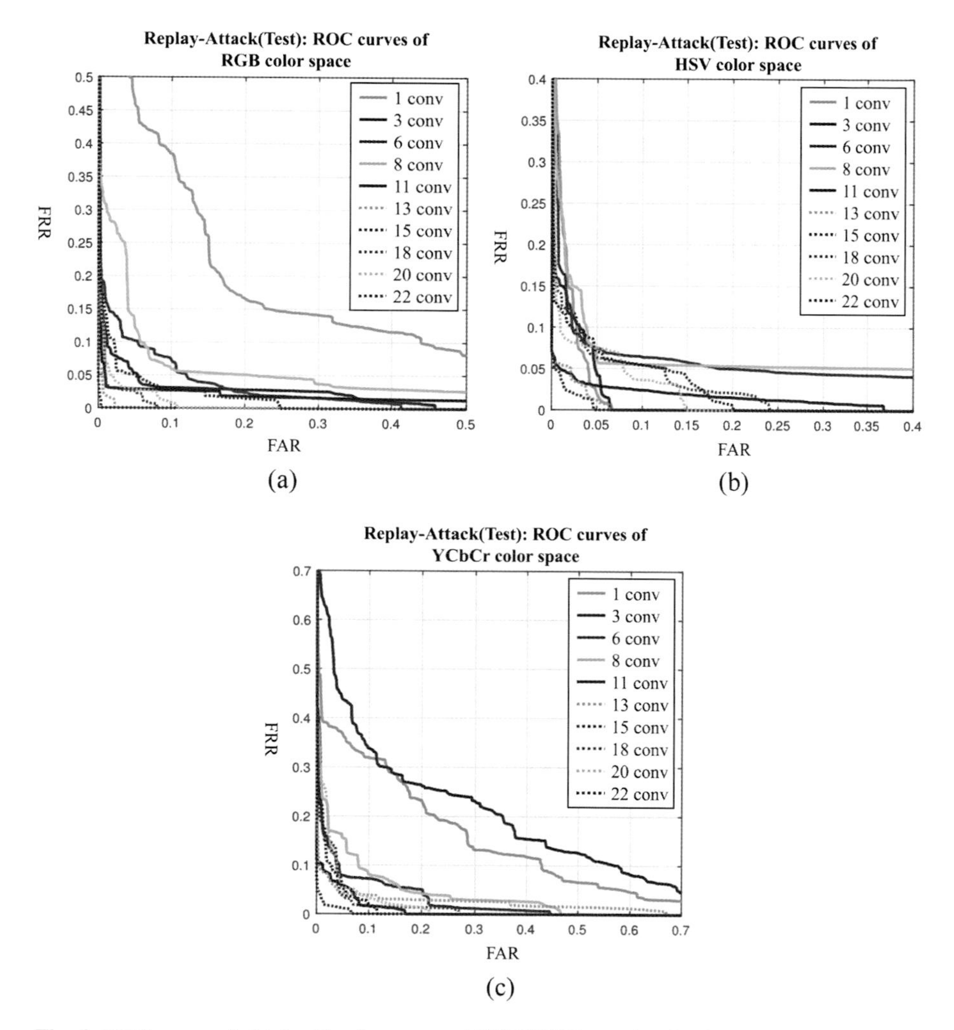

Fig. 4 ROC curves of obtained by the proposed CNN&LBP tested on Replay-Attack database. (**a**) ROC curves in RGB color space. (**b**) ROC curves in HSV color space. (**c**) ROC curves in YCbCr color space

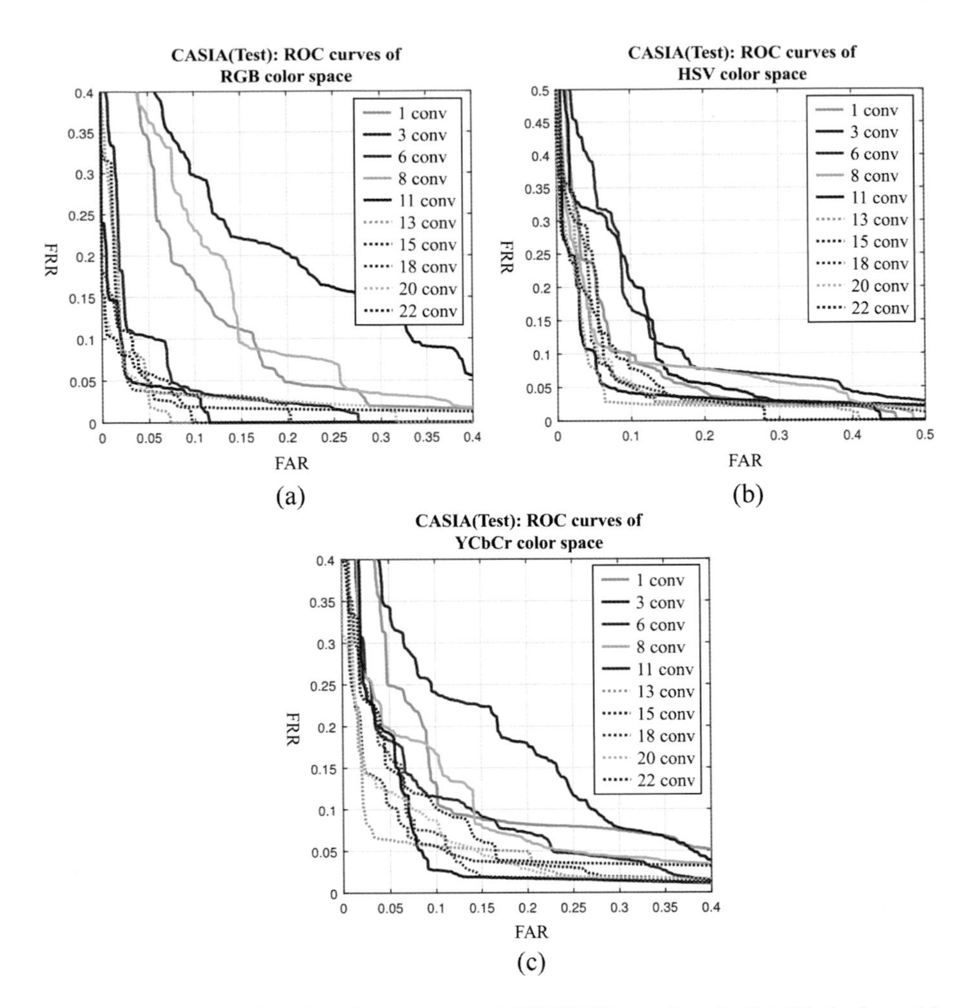

Fig. 5 ROC curves of obtained by the proposed CNN&LBP tested on CASIA-FA database. (**a**) ROC curves in RGB color space. (**b**) ROC curves in HSV color space. (**c**) ROC curves in HSV color space

distance to the original point is the blue dotted line. This means that for the Replay-Attack database, the 15th convolutional layer is the best layer to get the lowest EER and HTER. From both figures, we can conclude that the LBP descriptors extracted from the HSV color space have the strongest robustness in terms of all convolutional layers.

Table 3 The cross test results from different convolutional feature maps

Trained	Tested	Space	1 conv	3 conv	6 conv	8 conv	11 conv	13 conv	15 conv	18 conv	20 conv	22 conv
Replay-Attack	CASIA-FA	RGB	39.4	42.6	48.6	49.1	41.1	45.2	41.7	51.4	49.1	46.7
		HSV	**35.4**	46.2	48.5	37.3	43.9	40.3	42.9	47.4	39.3	40.8
		YCbCr	54.7	46.6	66.3	53.2	52.7	53.5	52.0	56.2	48.5	51.4
CASIA-FA	Replay-Attack	RGB	50.0	47.2	50.0	50.0	41.4	44.4	44.9	46.4	48.0	46.5
		HSV	38.6	42.7	40.1	**36.4**	38.1	36.7	40.1	42.5	40.0	38.9
		YCbCr	59.9	63.3	57.8	50.0	50.2	49.0	50.0	51.2	54.6	54.7

5.1.2 Cross Test

To evaluate the generalization capabilities of our proposed method, we first fine-tune the VGG-face model and train the SVM classifier on the Replay-Attack (or CASIA-FA) database and then test them on the CASIA-FA (or Replay-Attack) database. The results are shown in Table 3. It can be clearly seen that the LBP descriptors extracted from the HSV space have the strongest generalization capability. The lowest of HTERs for Replay-Attack and CASIA-FA are 35.4% and 36.4%, respectively. Comparing Table 2 with Table 3, the best convolutional layers for intra test are the 15th and the 18th, which are different with the 1_{st} and the 8th of the cross test. This suggests that the best LBP descriptors of one database maybe not suit for another database.

5.2 Concatenating Different Color Spaces

5.2.1 Intra Test

In order to take advantages of different color spaces, we concatenate the LBP features into one feature vector. More specifically, in our work, we explore to concatenate different color spaces with four mechanisms: RGB-HSV, RGB-YCbCr, HSV-YCbCr, and RGB-YCbCr-HSV. The detection results of intra test are shown in Table 4, and the ROC curves are illustrated in Figs. 6 and 7. Comparing Table 4 with Table 2, the overall performance has been significantly improved when concatenating the RGB and HSV color spaces. Especially the best EER of intra test has been reduced by more than 34%, even though the HTER of Replay-Attack database has been subtlely increased to 1.3%.

Figures 6 and 7 depict the robustness of the concatenated LBP descriptors. From the former, we can clearly see that all lines locate in the middle of the axis which means the concatenated LBP descriptors have weak robustness for Replay-Attack database. But for the CASIA database, the concatenated LBP descriptors have strong robustness illustrated in Fig. 7.

Table 4 The intra test results from different convolutional feature maps

Database			1 conv	3 conv	6 conv	8 conv	11 conv	13 conv	15 conv	18 conv	20 conv	22 conv
Replay-Attack	RGB-HSV	EER	4.0	4.0	5.2	6.6	2.6	1.5	**0.1**	2.8	2.8	2.8
		HTER	4.4	5.4	4.9	4.9	2.6	1.6	**1.3**	3.5	4.4	4.3
	RGB-YCbCr	EER	13.7	6.1	7.5	6.6	1.5	2.6	0.3	2.8	3.1	3.1
		HTER	14.0	8.1	7.3	6.2	1.8	2.5	1.5	3.9	3.8	3.7
	HSV-YCbCr	EER	5.9	6.6	5.0	5.3	1.2	2.6	0.5	3.5	3.3	4.0
		HTER	5.8	7.9	4.0	6.4	1.1	2.9	2.9	3.3	3.6	4.9
	RGB-HSV-YCbCr	EER	5.3	4.6	4.0	6.4	1.5	1.5	0.1	2.8	2.8	2.8
		HTER	6.1	6.2	4.8	6.8	1.4	2.0	1.5	2.9	2.9	4.0
CASIA-FA	RGB-HSV	EER	6.4	16.0	7.0	8.2	2.5	2.5	2.6	3.5	2.6	4.1
	RGB-YCbCr	EER	9.0	10.0	3.3	7.2	5.1	4.2	4.4	5.4	6.0	5.4
	HSV-YCbCr	EER	5.4	7.9	6.3	5.3	2.6	3.1	2.6	4.4	5.1	4.2
	RGB-HSV-YCbCr	EER	6.1	11.0	3.3	3.6	**2.3**	2.5	2.6	4.4	4.2	5.0

5.2.2 Cross Test

The generalization capabilities of the concatenated LBP descriptors are shown in Table 5. When our proposed method is trained on Replay-Attack and tested on CASIA-FA, the best HTER is 31.2%. Compared with Table 3, the performance of generalization has been improved by about 12%. Furthermore, when our proposed method is trained on CASIA-FA and tested on Replay-Attack, the generalization capabilities will be significantly improved. Especially for the concatenated RGB-HSV color spaces, the best HTER can be decreased from 36.4% to 20.7%. Tables 3 and 5 show the concatenated color spaces can improve the generalization capabilities for face anti-spoofing.

5.3 Comparison Against State-of-the-Art Methods

Tables 6 and 7 provide a comparison with the state-of-the-art methods. It can be seen from the two tables that our proposed method outperforms the state-of-the-art algorithms on Replay-Attack and CASIA-FA. More specially, Table 6 compares the intra test performance with other methods. Considering Replay-Attack database, we find that the EER obtained by the proposed method is 0.1%, which is very close to the best EER reported in [8]. For the indicator of HTER, our performance is better nearly four times than [8]. Especially for CASIA-FA, the maximum of EER is 32.4%, which is more than 14 times than ours. Table 7 compares the generalization

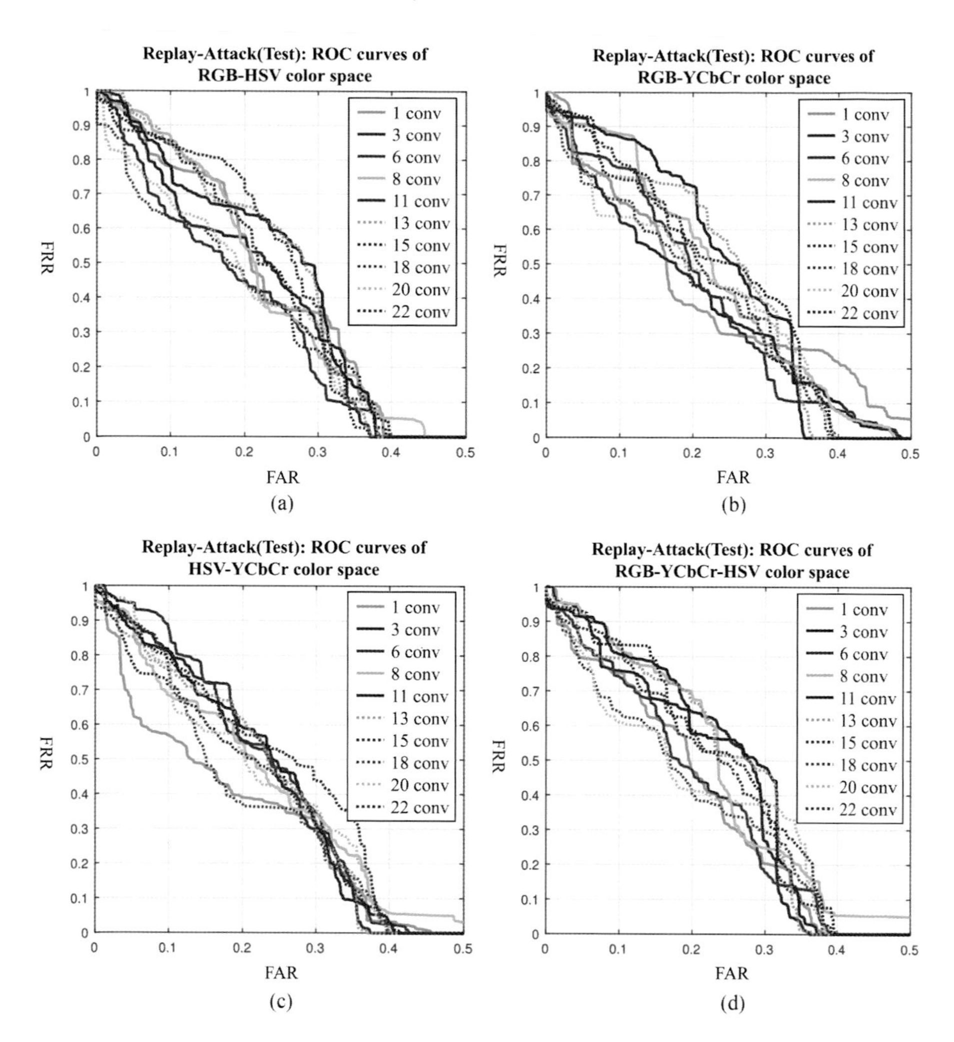

Fig. 6 ROC curves of Replay-Attack database obtained by the concatenated color spaces. (**a**) The RGB-HSV space. (**b**) The RGB-YCbCr space. (**c**) The HSV-YCbCr space. (**d**) The RGB-YCbCr-HSV space

capabilities with the state-of-the-art methods. When the our proposed method is trained on the CASIA-FA, we notice that the HTER value is 20.7%. When the model is trained on Replay-Attack, the HTER on CASIA-FA is 31.2%. So overall, our proposed approach gives the competitive performance on the challenging Replay-Attack and CASIA-FA.

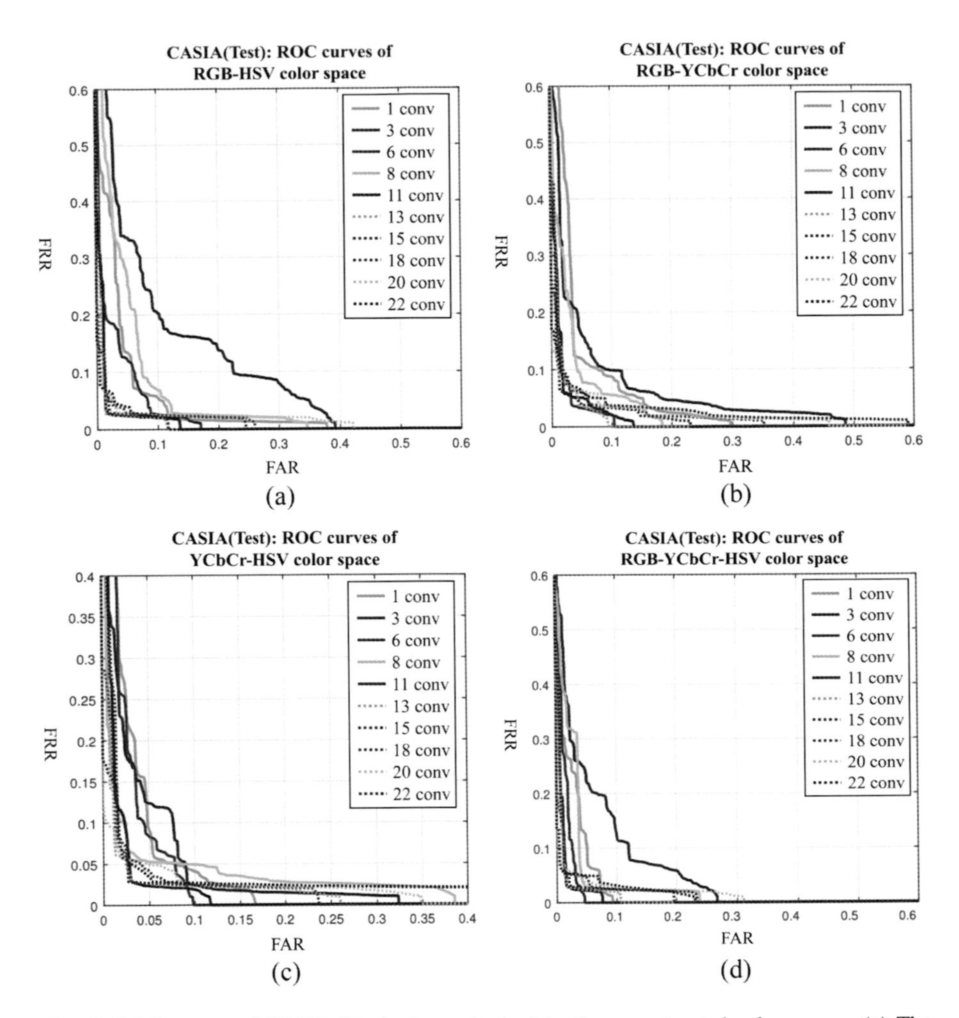

Fig. 7 ROC curves of CASIA-FA database obtained by the concatenated color spaces. (**a**) The RGB-HSV space. (**b**) The RGB-YCbCr space. (**c**) The HSV-YCbCr space. (**d**) The RGB-YCbCr-HSV space

6 Conclusion

In this chapter, we propose to detect face spoofing attacks based on deep color texture analysis. We extract color texture features from convolutional feature maps and investigate how well different color image representations (RGB, HSV, and YCbCr) can be used for describing the intrinsic disparities in the color textures between genuine faces and fake ones and if they can provide complementary representations. The effectiveness of different deep color texture representations is also studied by concatenating different color spaces. Extensive experiments on

Table 5 The cross test results from different convolutional feature maps

Trained	Tested	Space	1 conv	3 conv	6 conv	8 conv	11 conv	13 conv	15 conv	18 conv	20 conv	22 conv
Replay-Attack	CASIA-FA	RGB-HSV	33.5	47.3	40.8	44.3	40.3	37.3	42.1	42.2	41.1	41.2
		RGB-YCbCr	46.9	45.7	54.4	48.5	47.4	46.6	47.8	52.8	45.3	44.6
		HSV-YCbCr	**31.2**	44.9	51.2	39.1	45.4	44.2	37.9	45.6	39.5	37.2
		RGB-HSV-YCbCr	32.4	47.7	46.5	41.4	41.5	39.1	43.1	40.6	42.6	35.8
CASIA-FA	Replay-Attack	RGB-HSV	47.5	41.2	44.5	43.0	28.9	23.2	27.9	27.2	21.9	**20.7**
		RGB-YCbCr	50.9	57.0	49.8	50.0	47.5	46.7	48.9	47.7	50.8	49.4
		HSV-YCbCr	52.1	37.9	47.8	42.8	45.1	45.6	42.8	38.5	39.2	40.2
		RGB-HSV-YCbCr	49.9	40.1	49.9	45.9	38.8	41.4	41.3	35.9	37.8	38.1

Table 6 Compare the performance with the state-of-the-art methods

Method	Replay-Attack		CASIA-FA
	EER(%)	HTER(%)	EER(%)
IQA based [18]	–	–	32.4
CDD [32]	–	–	11.8
DOG(baseline) [30]	–	–	17.0
Motion+LBP [33]	4.5	5.1	–
Motion [31]	11.6	11.7	26.6
LBP [29]	13.9	13.8	18.2
LBP-TOP [13]	7.8	7.6	10.6
Spectral cubes [34]	–	2.8	14.0
DMD [35]	5.3	3.8	21.8
Deep Learning [9]	6.1	2.1	7.3
LBP [36]	0.4	2.9	6.2
Color LBP [8]	**0.0**	3.5	3.2
Proposed method	0.1	**0.9**	**2.3**

the two latest and challenging face spoofing databases (Replay-Attack database and CASIA-FA database) show excellent results. More importantly, our proposed method is able to achieve stable performance across the two databases unlike most of the methods proposed in the literature. Furthermore, in our cross test, the deep color texture representation showed promising generalization capabilities, thus suggesting that our method seems to be more stable in unknown conditions compared to the state-of-the-art methods.

As a future work, we plan to evaluate our methodology on other kinds of anti-spoofing tasks such as fingerprint and Iris. It is finally worth noting that one can expect much better results with the availability of training images. So, we will set up a more large-scale database for face anti-spoofing.

Table 7 Compare the cross test results with the state-of-arts

Method	Trained	Tested	HTER
Motion [31]	CASIA-FA	Replay-Attack	50.2
	Replat-Attack	CASIA	47.9
LBP [31]	CASIA-FA	Replay-Attack	45.9
	Replay-Attack	CASIA-FA	57.6
LBP-TOP [31]	CASIA-FA	Replay-Attack	47.9
	Replay-Attack	CASIA-FA	60.6
Spectral cubes [34]	CASIA-FA	Replay-Attack	34.4
	Replay-Attack	CASIA-FA	50.0
Deep Learning [9]	CASIA-FA	Replay-Attack	48.5
	Replay-Attack	CASIA-FA	45.5
LBP [36]	CASIA-FA	Replay-Attack	47.0
	Replay-Attack	CASIA-FA	39.6
Color LBP [8]	CASIA-FA	Replay-Attack	30.3
	Replay-Attack	CASIA-FA	37.7
Proposed method	CASIA-FA	Replay-Attack	**20.7**
	Replay-Attack	CASIA-FA	**31.2**

References

1. Y. Li, K. Xu, Q. Yan, Y. Li, R.H. Deng, Proceedings of Acm Sigsac Symposium on Information Computer and Communications Security (2014)
2. L. Li, X. Feng, X. Jiang, Z. Xia, A. Hadid, in *IEEE International Conference on Image Processing (ICIP)* (2017), pp. 101–105. DOI 10.1109/ICIP.2017.8296251
3. L. Li, X. Feng, Z. Boulkenafet, Z. Xia, M. Li, A. Hadid, in *2016 Sixth International Conference on Image Processing Theory, Tools and Applications (IPTA)* (2016), pp. 1–6. DOI 10.1109/IPTA.2016.7821013
4. Z. Boulkenafet, J. Komulainen, A. Hadid, in *Image Processing (ICIP), 2015 IEEE International Conference on* (2015), pp. 2636–2640. DOI 10.1109/ICIP.2015.7351280
5. D. Wen, H. Han, A.K. Jain, IEEE Transactions on Information Forensics and Security **10**(4), 746 (2015)
6. J. Maatta, A. Hadid, M. Pietikainen, in *International Joint Conference on Biometrics* (2011), pp. 1–7
7. I. Chingovska, A. Anjos, S. Marcel, in *IEEE International Conference of the Biometrics Special Interest Group* (2012), pp. 1–7
8. Z. Boulkenafet, J. Komulainen, A. Hadid, IEEE Transactions on Information Forensics and Security **11**(8), 1 (2016)
9. J. Yang, Z. Lei, S.Z. Li, Eprint Arxiv **9218**, 373 (2014)
10. D. Menotti, G. Chiachia, A. Pinto, W. Robson Schwartz, H. Pedrini, A. Xavier Falcao, A. Rocha, Computer Science **10**, 864 (2014)
11. A. Krizhevsky, I. Sutskever, G.E. Hinton, Advances in Neural Information Processing Systems **25**(2), 2012 (2012)
12. A. Anjos, S. Marcel, in *Biometrics, International Joint Conference on* (2011), pp. 1–7
13. T.D.F. Pereira, A. Anjos, J.M.D. Martino, S. Marcel, in *Proceedings of the 11th international conference on Computer Vision - Volume Part I* (2012), pp. 121–132
14. K. Kollreider, H. Fronthaler, J. Bigun, Image and Vision Computing **27**(3), 233 (2009)

15. M.M. Chakka, A. Anjos, S. Marcel, R. Tronci, in *Proceedings of the 2011 International Joint Conference on Biometrics* (2011), pp. 1–6
16. Z. Zhang, J. Yan, S. Liu, Z. Lei, in *Biometrics (ICB), 2012 5th IAPR International Conference on* (2012), pp. 26–31
17. J. Peng, P.P.K. Chan, in *Wavelet Analysis and Pattern Recognition (ICWAPR), 2014 International Conference on* (2014), pp. 176–181. DOI 10.1109/ICWAPR.2014.6961311
18. J. Galbally, S. Marcel, in *International Conference on Pattern Recognition (ICPR)* (2014), pp. 1173–1178. DOI 10.1109/ICPR.2014.211
19. Z. Zhang, D. Yi, Z. Lei, S.Z. Li, in *IEEE International Conference on Automatic Face and Gesture Recognition and Workshops* (2011), pp. 436–441
20. G. Chetty, **23**(3), 1 (2010)
21. T. Wang, J. Yang, Z. Lei, S. Liao, in *International Conference on Biometrics* (2013), pp. 1–6
22. Z. Akhtar, C. Michelon, G.L. Foresti, in *Security Technology (ICCST), 2014 International Carnahan Conference on* (2014), pp. 6695–6695
23. Z. Akhtar, C. Micheloni, C. Piciarelli, G.L. Foresti, in *Advanced Video and Signal Based Surveillance (AVSS), 2014 11th IEEE International Conference on* (2014), pp. 187–192. DOI 10.1109/AVSS.2014.6918666
24. D. Menotti, G. Chiachia, A. Pinto, W. Robson Schwartz, H. Pedrini, A. Xavier Falcao, A. Rocha, Information Forensics and Security, IEEE Transactions on **10**(4), 864 (2015)
25. O.M. Parkhi, A. Vedaldi, A. Zisserman, in *British Machine Vision Conference* (2015)
26. R. Lukac, K.N. Plataniotis, *Color image processing: methods and applications* (CRC, Taylor and Francis, 2007)
27. T. Ojala, M. Pietikainen, T. Maenpaa, IEEE Transactions on Pattern Analysis and Machine Intelligence **24**(7), 971 (2002)
28. R.E. Fan, K.W. Chang, C.J. Hsieh, X.R. Wang, C.J. Lin, Journal of Machine Learning Research **9**(12), 1871 (2010)
29. I. Chingovska, A. Anjos, S. Marcel, in *International Conference of the Biometrics Special Interest Group (BIOSIG)* (2012), pp. 1–7
30. Z. Zhang, J. Yan, S. Liu, Z. Lei, D. Yi, S. Li, in *International Conference on Biometrics (ICB)* (2012), pp. 26–31
31. T. de Freitas Pereira, A. Anjos, J. De Martino, S. Marcel, in *International Conference on Biometrics (ICB)* (2013), pp. 1–8
32. J. Yang, Z. Lei, S. Liao, S. Li, in *Biometrics (ICB), 2013 International Conference on* (2013), pp. 1–6
33. J. Komulainen, A. Hadid, M. Pietikainen, A. Anjos, S. Marcel, in *International Conference on Biometrics (ICB)* (2013), pp. 1–7. DOI 10.1109/ICB.2013.6612968
34. A. Pinto, H. Pedrini, W.R. Schwartz, A. Rocha, IEEE Transactions on Image Processing A Publication of the IEEE Signal Processing Society **24**(12), 4726 (2015)
35. S. Tirunagari, N. Poh, D. Windridge, A. Iorliam, N. Suki, A.T.S. Ho, IEEE Transactions on Information Forensics and Security **10**(4), 762 (2015)
36. Z. Boulkenafet, J. Komulainen, A. Hadid, in *IEEE International Conference on Image Processing (ICIP2015)* (2015)

Kinship Verification Based on Deep Learning

Xiaoting Wu, Xiaoyi Feng, Lei Li, Elhocine Boutellaa, and Abdenour Hadid

Abstract Automatic kinship verification using facial images is a relatively new and challenging research problem in computer vision. It consists in automatically predicting whether two persons have a biological kin relation by examining their facial attributes. While most of the existing works extract shallow handcrafted features from still face images, in this chapter, we approach this problem from deep learning point of view. Promising results, especially those of deep features, are obtained on the benchmark UvA-NEMO Smile database and KinFaceW-I and KinFaceW-II kinship face databases.

1 Introduction

It is a common and an easy practice for us, humans, to identify our relatives from faces. Relatives usually wonder which facial attributes does a new born baby inherit from which family member. The human ability of kinship recognition has been the object of many psychological studies [8, 9]. Inspired by these studies, automatic kinship (or family) verification [12, 37] has been recently considered as an interesting and open research problem in computer vision which is receiving an increasing attention from the research community.

Automatic kinship verification from faces aims at determining whether two persons have a biological kin relation or not by comparing their facial attributes. Kinship verification is important for automatically analyzing the huge amount of photos daily shared on social media. It helps understanding the family relationships in these photos. Kinship verification is also useful in case of missing children or elderly people with Alzheimer or possible kidnapping cases. For instance, a

X. Wu · X. Feng (✉) · L. Li
School of Electronics and Information, Northwestern Polytechnical University, Xi'an, China
e-mail: wuxt14@mail.nwpu.edu.cn; fengxiao@nwpu.edu.cn; li_lei_08@163.com

E. Boutellaa · A. Hadid
Center for Machine Vision and Signal Analysis, University of Oulu, Oulu, Finland
e-mail: Elhocine.Boutellaa@oulu.fi; hadid@ee.oulu.fi

X. Jiang et al. (eds.), *Deep Learning in Object Detection and Recognition*,
https://doi.org/10.1007/978-981-10-5152-4_5

suspicious behavior between two persons (e.g., an adult and a child) captured by a surveillance camera can be subjected to further analysis to determine whether they are from the same family or not to prevent crimes and kidnapping. Kinship verification can also be used for automatically organizing family albums and generating family trees.

Kinship verification using facial images/videos is a very challenging task. It inherits the research problems of face verification from images captured in the wild under adverse pose, expression, illumination, and occlusion conditions. In addition, kinship verification should deal with wider intra-class and inter-class variations, as persons from the same family may look very different while faces of persons with no kin relation may look similar. Moreover, automatic kinship verification poses new challenges, since a pair of input images may be from persons of different sexes (e.g., brother-sister kin) and/or with a large age difference (e.g., father-daughter kin).

The published papers and organized competitions (e.g., [21, 24]) dealing with automatic kinship verification over the past few years have shown some promising results. Typical current best-performing methods combine several face descriptors, apply metric learning approaches, and compute Euclidean distances between pairs of features for kinship verification. It appears that most of these works are mainly based on shallow handcrafted features. Hence, they are not associated with the recent significant progress in machine learning, which suggests the use of deep features. In this present work, we introduce novel approaches to kinship verification from facial images using deep learning from both static and dynamic points of view and to exploit the recent progress in deep learning for facial analysis.

The rest of the chapter is organized as follows. Section 2 discusses the related work. Image-based kinship verification and video-based kinship verification are described in Sects. 3 and 4, respectively. Section 5 draws conclusions and points out future research directions.

2 Related Work

Quite many research papers dealing with kinship verification from face have already been published. Moreover, two kinship verification competitions have been held in the past years. The first competition was held in 2014 in conjunction with the International Joint Conference on Biometrics (IJCB'2014), Clearwater, Florida, USA [22]. Four participants took part in this first competition. The second competition was held in conjunction with the 2015 International Conference on Automatic Face and Gesture Recognition (FG 2015), Ljubljana, Slovenia [23]. There were five teams participating to this last competition. In both competitions, the KinFaceW (I and II) database and protocols are used to evaluate the competing algorithms. A remarkable improvement of verification performances is noticed in the latter competition compared to the former. However, in the second competition, the test set was available to the participants, while it was not the case for the first competition. In the following, we present a summary of the literature of kinship verification from face images. According to the main contribution, we group the proposed approaches into three different categories:

2.1 Methods Based on Features

Face description is a key step in any face analysis system. Indeed, if the extracted face features are enough discriminative, the classification can be simply performed with a linear classifier such as nearest neighbor. For kinship problem, different feature extraction methods [22, 23, 30, 32, 38] and strategies, including selection [13], learning [34], and fusion [4], have been investigated. The first kinship works focused on selecting useful features. For instance, the authors of [13] selected approximately 22 kinds of facial features. Through the experiments, they rank the features according to their individual performances and choose top 14 features, which achieved an accuracy rate of more than 50%. The selected features can be divided into two types: skin color features and distances between different face parts. Both appearance and geometry feature have been used by Wang et al. [30]. As appearance features, pyramid images are built for each face by considering different sizes, and then LBP features were extracted from overlapping patches. A Gaussian mixture model is applied to find the corresponding similar patch pairs between two face images. The appeared feature is taken as the absolute difference between two GMM patch vectors. As for geometric features, facial landmarks are first detected and then projected to the Grassmann manifold, and the geodesic distance is used to measure the difference of two face shapes. Yan et al. [34] learned mid-level features correspond to decision values from support vector machine (SVM) hyperplanes. An objective function is formulated on the learned features so that face samples with a kin relation are expected to have similar decision values from the hyperplanes. For training, a large unlabeled face dataset and a small dataset of face pairs labeled with kinship relations are used. The authors further introduced a multi-view method to learn a common mid-level feature representation from multiple low-level descriptors. The used low-level features are LBP, spatial pyramid learning (SPLE), and SIFT. Feature fusion has been the aim of [4], where four different textural features are extracted: LPQ, WLD, TPLBP, and FPLBP. For each feature, the difference between vectors of a pair images is computed and normalized, and the four features are concatenated forming the pair descriptor. The minimum redundancy maximum relevance algorithm is applied to perform feature selection, and SVM was utilized for classification. This approach obtained the best performance on the restrict image configuration in the second kinship competition [23].

2.2 Methods Using Metric Learning

Since facial appearance similarity between parents and children is grater that it is between unrelated persons, a particular interest has been attributed to the use of different metric learning approaches for solving the kinship verification problem. Metric learning aims at automatically learning a similarity measure from data rather than using handcrafted distances. In the kinship verification problem, the aim is to learn a metric where the distance between a face feature pair with a kin relation is smaller than it is between pairs with no relation. Among the first approaches to tackle kinship verification, Somanath and Kambhamettu [29] applied ensemble metric learning. The training data is initially clustered into using different similarity kernels. Then a final kernel is learned based on the initial clustering. For each kin relation, the learned kernel ensures that related pairs have a greater similarity than unrelated pairs. Similar to the previous work, recently Zhou et al. [40] applied ensemble similarity learning for solving the kinship verification problem. They learned an ensemble of sparse bilinear similarity bases from kinship data by minimizing the violation of the kinship constraints between pairs of images and maximizing the diversity of the similarity bases. Lu et al. [20] learned a distance metric where the face pairs with a kin relation are pulled close to each other and those without a kin relation are pushed away. Yan et al. [33] and Hu et al. [15] learned multiple distance metrics based on various features, by simultaneously maximizing the kinship constraint (pairs with a kinship relation must have a smaller distance than pairs without a kinship relation) and the correlation of different features.

2.3 Other Methods

Among the other approaches for kinship verification, Xia et al. [31] applied subspace transfer learning on UB KinFace database to mitigate the huge divergence between children and old parents. To this end, young parents are considered as an intermediate domain between children and old parents. Transfer learning was achieved by seeking a subspace projection where the intermediate domain (young parents) and the source (children) and target (old parents) domains share the same distribution.

Recently, Chen et al. [6] approached the kinship verification problem from a multi-linear coherent spaces perspective where parent and child images are considered as two different views data. Then, local image patches at different scales are independently projected into coherent spaces via canonical correlation analysis such that patch pairs with kinship relations are more correlated. Further, only useful face patches for kinship verification were selected.

Instead of verifying kinship from pairs of face images, Qin et al. [28] investigated tri-subject kinship verification. Two types of relations are considered: father-mother-daughter and father-mother-son. The authors introduced a relative symmetric bilinear approach to model the similarity between the child and the parents (one-versus-two matching).

Kou et al. [18] proposed an online sparse similarity learning approach for kinship verification. The authors formulated two kinship triplet constraints for a quadruple input, involving a positive pair and a negative pair, so that similarity of pairs with a kin relation is higher than that of pairs from different families.

3 Image-Based Kinship Verification

In this work, we propose a novel approach to kinship verification from facial images using similarity metric-based convolutional neural networks (SMCNN). This approach is partially inspired by the work of Yan et al. [33] which proposed a metric learning method based on deep neural networks for the task of face verification. However, different from the work of Yan et al. which considered handcrafted features as inputs to the deep neural network, our approach explores automatically learned features for kinship verification.

3.1 Methodology

As can be seen from Fig. 1, the SMCNN architecture consists of two identical convolutional neural networks. Each of them contains *eight* layers (see Fig. 2):

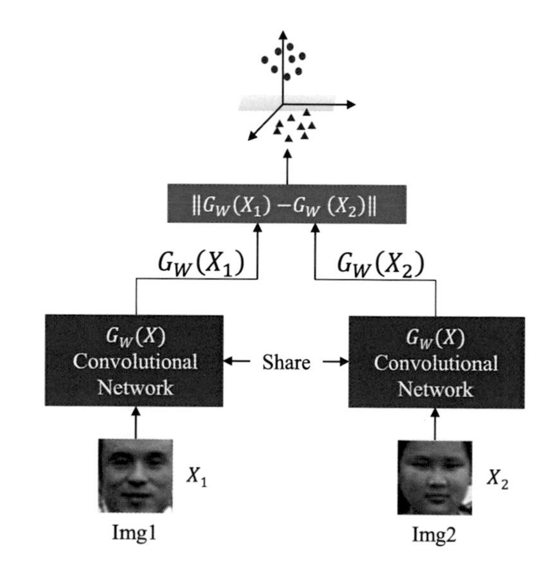

Fig. 1 The framework of our proposed approach using similarity metric-based convolutional neural networks (SMCNN) for kinship verification. An image pair is first fed into two identical convolutional neural networks. The L_1 norm between their outputs is then computed. Finally, a decision is made based on a learned threshold

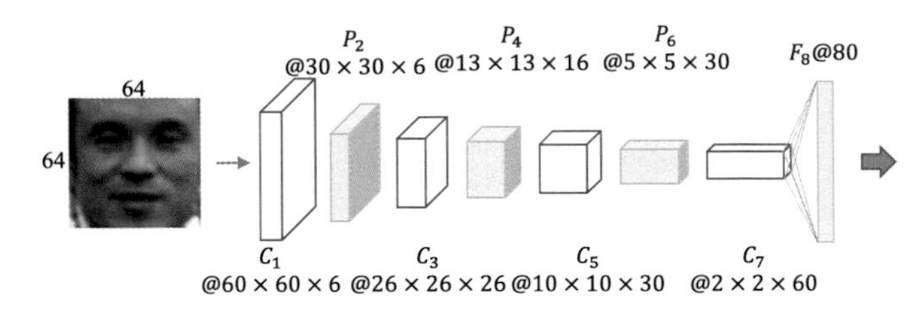

Fig. 2 The architecture of the convolutional neural network in SMCNN. The dark green represents the convolutional layers, shallow green represents the subsampling layers, and lastly blue layer is the full connection layer

Table 1 Parameters of the convolutional neural network

Layers	C_1	P_2	C_3	P_4	C_5	P_6	C_7	F_8
Input-num	1	6	6	16	16	30	30	60
Out-num	6	6	16	16	30	30	60	80
Filter-size	5 × 5	2 × 2	5 × 5	2 × 2	4 × 4	2 × 2	4 × 4	2 × 2

four convolutional layers, *three* pooling layers, and a full connection layer. The connection orders of these layers are as follows: $C_1 - P_2 - C_3 - P_4 - C_5 - P_6 - C_7 - F_8$, where C_t is the convolutional layer, P_t denotes a pooling (subsampling) layer, and F_t represents a full connection layer. t is the layer index. In our proposed approach, the pooling layer is carrying the subsampling operation. The detailed parameters of each layer are given in Table 1.

Let x_i and x_j denote the grayscale input images (i.e., corresponding to the grayscale images of $Img1$ and $Img2$ in Fig. 1) and y the label of the image pair (x_i, x_j). $y = 1$ if the image pair has a kinship relationship (referred to as positive pair) and $y = -1$ otherwise (referred to as negative pair). W represents the weights of the convolutional neural network. $G_W(x_i)$ and $G_W(x_j)$ represent the features given by the convolutional neural network. The distance between $G_W(x_i)$ and $G_W(x_j)$ is measured by L_1 norm, which is defined as:

$$D_G(x_i, x_j) = \|G_W(x_i) - G_W(x_j)\|_1 \tag{1}$$

The distance $D_G(x_i, x_j)$ should be smaller for positive image pairs and larger for negative image pairs. In [16], two thresholds have been used for classification. If $D_G(x_i, x_j)$ is smaller than a predefined threshold τ_1, then $D_G(x_i, x_j)$ belongs to a positive pair. Otherwise, if $D_G(x_i, x_j)$ is larger than a threshold $\tau_2 (\tau_1 < \tau_2)$, then $D_G(x_i, x_j)$ belongs to a negative pair.

For simplicity of the classification scheme using the two convolutional neural networks, the thresholds τ_1 and τ_2 can be combined into one threshold τ so that τ is larger than 1 and:

$$y(\tau - D_G(x_i, x_j)) > 1 \tag{2}$$

where $\tau_1 = \tau - 1$ and $\tau_2 = \tau + 1$. The cost function of SMCNN can be formulated as follows:

$$\begin{aligned} \underset{G}{argmin} J &= J_1 + J_2 \\ &= \sum_{i,j} f(1 - y(\tau - D_G(x_i, x_j))) \\ &+ \frac{\lambda}{2}(\|W\|_{\mathbb{F}}^2 + \|b\|_{\mathbb{F}}^2) \end{aligned} \tag{3}$$

where $f(t) = \frac{1}{\beta} log(1 + exp(\beta t))$ is the generalized logistic loss [25], β is a sharpness parameter, operation $\|M\|_{\mathbb{F}}$ denotes the Frobenius norm of the matrix M, and λ is a regularization parameter.

To minimize the cost function in Eq. 3, sub-gradient descent algorithm is adopted in the convolutional neural network. W_t denotes the weighting parameters in t layer, and b_t represents the bias term. Based on the principle of backpropagation, the key idea is to calculate the error term δ_F in the full connection layer. So, the error term of SMCNN is:

$$\delta = y \cdot f'(u) \cdot sign(c) \tag{4}$$

where y is the label of the image pair ($y = 1$ or $y = -1$), $u = 1 - y(-D_G(x_i, x_j))$, $c = abs(G_W(x_i) - G_W(x_j))$, and $sign$ is the sign function which is an approximation of the partial derivative c in abs function, as illustrated in Eq. 5:

$$sign(c) = \begin{cases} -1, & c < 0 \\ 0, & c = 0 \\ +1, & c > 0 \end{cases} \tag{5}$$

The error term of the full connection layer in the convolutional neural network can be calculated as follows:

$$\begin{aligned} \delta_{F_i} &= \delta \odot s'(F_8(x_i)) \\ \delta_{F_j} &= -\delta \odot s'(F_8(x_j)) \end{aligned} \tag{6}$$

where $F_8(x_i)$ is the output of the full connection layer and the symbol s is the activation function of the full connection layer. After computing the error term δ_F of the full connection layer, the partial derivative of W_t and b_t can be computed by backpropagation [7]. Then, gradient descent algorithm can be used to update the corresponding parameters as follows:

$$
\begin{aligned}
W_t &= W_t - \rho \cdot \frac{\partial J}{\partial W} \\
b_t &= b_t - \rho \cdot \frac{\partial J}{\partial b}
\end{aligned} \tag{7}
$$

where ρ is a learning rate which is set in our experiments to $\rho = 0.001$.

Used Toolbox We used the deep learning toolbox called "DeepLearnToolbox" [26]. DeepLearnToolbox does not contain a ready framework for SMCNN. So, we made changes to build the SMCNN. The changes are mainly related to two aspects. First, we changed the connection between the first convolutional layer and the second convolutional layer. Second, we changed the activation function from $sigmoid$ to $tanh$.

Partial Connection To extract different (hopefully complementary) features and keep the number of connections within reasonable bounds, the partial connection scheme is adopted between the first convolutional layer and the third convolutional layer, as illustrated in Table 2.

Activation Function We use the $tanh$ function as the activation function in SMCNN. The $tanh$ function and its derivative are computed as follows:

$$
\begin{aligned}
s(z) &= tanh(z) = \frac{exp(z) - exp(-z)}{exp(z) + exp(-z)} \\
s'(z) &= tanh'(z) = 1 - tanh^2(z)
\end{aligned} \tag{8}
$$

Parameter Initialization We randomly initialized the weights W_t and the bias term b_t based on $Gaussian$ distribution, with mean value of 0 and standard deviation of 0.05.

Table 2 Connection scheme between layers C_1 and C_3

		C3															
		1	2	3	4	5	6	7	8	9	10	11	12	13	14	15	16
C1	1	√				√	√	√			√	√	√	√		√	√
	2	√	√				√	√	√			√	√	√	√		√
	3	√	√	√				√	√	√			√		√	√	√
	4		√	√	√			√	√	√	√			√		√	√
	5			√	√	√			√	√	√	√		√	√		√
	6				√	√	√			√	√	√	√		√	√	√

3.2 Experimental Analysis

To evaluate the performance of our proposed approach SMCNN, we experimented with the publicly available benchmark datasets KinFaceW [19] which can be seen in Fig. 3. KinFaceW datasets contain KinFaceW-I as well as KinFaceW-II and include four kinship relationships: father-son (FS), father-daughter(FD), mother-son (MS), and mother-daughter (MD). The number of kinship pair relations in KinFaceW-I is 156, 134, 116, and 127 for FS, FD, MS, and MD, respectively. In KinFaceW-II, there are in total 250 kinship pair relations. Based on the webpage of KinFaceW, we use the pre-specified training/testing split, which was generated randomly and independently for five folds. Four folds are used for training, while the remaining fold is used for testing.

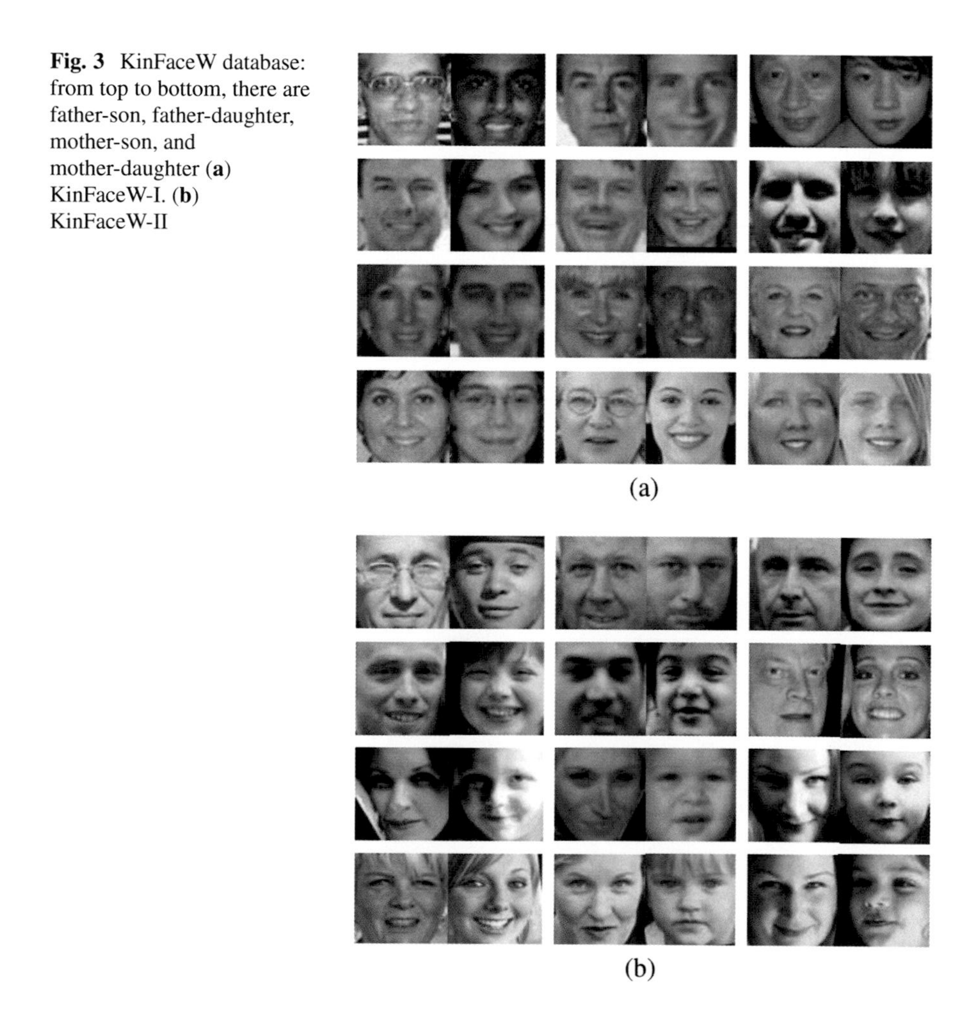

Fig. 3 KinFaceW database: from top to bottom, there are father-son, father-daughter, mother-son, and mother-daughter (**a**) KinFaceW-I. (**b**) KinFaceW-II

Table 3 Compare the fine-tuning results with the others (accuracy: %)

Method	KinFaceW-I				KinFaceW-II			
	FD	FS	MD	MS	FD	FS	MD	MS
DMML2014 [33]	74.5	69.5	69.5	**75.5**	78.5	**76.5**	78.5	**79.5**
MNRML2014 [20]	72.5	66.5	66.2	72.0	76.9	74.3	77.4	77.6
SM 2015 [18]	66.1	62.2	64.3	70.0	74.9	71.0	76.9	76.4
Proposed SMCNN	**75.0**	**75.0**	**72.2**	68.7	**79.0**	75.0	**85.0**	78.0

As can be seen from Table 3, our proposed approach compares favorably against the related methods in the state of the art. On KinFaceW-I, our approach yields in the best performance for all kinship relationships except for MS relationship for which the best results are obtained with DMML 2014 [33]. On KinFaceW-II, our proposed approach gives the best results in two cases (FD and MD), while DMML 2014 [33] yields in slightly better results in two other cases (FS and MS). Note that our approach outperforms MNRML 2014 [20] and SM 2015 [18] methods in all cases and for both KinFaceW-I and KinFaceW-II.

We can notice from Table 3 that the results on KinFaceW-II are better than those on KinFaceW-I. The reason is that the images in KinFaceW-II are cropped from the same pictures and hence sharing similar environment, such as illuminate intensity and chrominance. However, KinFaceW-I is collected from different pictures in uncontrolled environments.

4 Video-Based Kinship Verification

The published papers and organized competitions (e.g., [21, 24]) dealing with automatic kinship verification over the past few years have shown some promising results. Typical current best-performing methods combine several face descriptors, apply metric learning approaches, and compute Euclidean distances between pairs of features for kinship verification. It appears that most of these works are mainly based on shallow handcrafted features. Hence, they are not associated with the recent significant progress in machine learning, which suggests the use of deep features. Moreover, the role of facial dynamics in kinship verification is mostly unexplored as almost all the existing works focus on analyzing still facial images instead of video sequences. Based on these observations, we propose to approach the problem of kinship verification from a spatiotemporal point of view and to exploit the recent progress in deep learning for facial analysis.

Given two face video sequences, to verify their kin relationship, our proposed approach starts with detecting, segmenting, and aligning the face images based on eye coordinates. Then, two types of descriptors are extracted: shallow spatiotemporal texture features and deep features. As spatiotemporal features, we extract local binary patterns (LBP) [1], local phase quantization (LPQ) [2], and binarized statistical image features (BSIF) [17]. These features are all extracted from three orthogonal planes (TOP) of the videos. Deep features are extracted by convolutional neural networks (CNNs) [27]. The feature vectors of face pairs to compare are then combined to be used as inputs to support vector machines (SVM) for classification. We conduct extensive experiments on the benchmark UvA-NEMO Smile database [10] obtaining very promising results, especially with the deep features. The results also clearly demonstrate the superiority of using videos over still images, hence pointing out the important role of facial dynamics in kinship verification. Furthermore, the fusion of the two types of features (i.e., shallow spatiotemporal texture features and deep features) results in significant performance improvements compared to state-of-the-art methods.

4.1 Methodology

In our approach, the first step consists in segmenting the face region from each video sequence. For that purpose, we have employed an active shape model (ASM)-based approach that detects 68 facial landmarks. The regions containing faces are then cropped from every frame in the video using the detected landmarks. Finally, the face regions are aligned using key landmark points and registered to a predefined template.

For describing faces from videos, we use two types of features: texture spatiotemporal features and deep learning features. These features are introduced in this subsection. *Spatiotemporal features* Spatiotemporal texture features have been shown to be efficient for describing faces in various face analysis tasks, such as face recognition and facial expression classification. In this work, we extract three local texture descriptors: LBP [1], LPQ [2], and BSIF [17]. These three features are able to describe an image using a histogram of decimal values. The code corresponding to each pixel in the image is computed from a series of binary responses of the pixel neighborhood to a filter bank. In LBP and LPQ, the filters are handcrafted, while the filters of BSIF are learned from natural images. Specifically, the binary code of a pixel in LBP is computed by thresholding its value with the circularly symmetric P neighboring pixels (on a circle of radius R). LPQ encodes the local phase information of four frequencies of the short-term Fourier transform (STFT) over a local window of size $W \times W$ surrounding the pixel. BSIF binarizes the responses of f independent filters of size $W \times W$ learnt by independent component analysis (ICA).

The spatiotemporal textural dynamics of the face in a video are extracted from three orthogonal planes XY, XT, and YT [36], separately. X and Y are the horizontal and vertical spatial axes of the video, and T refers to the time. The texture features of each plane are aggregated into a separate histogram. Then the three histograms are concatenated into a single feature vector. To take benefit of the multi-resolution representation [5], the three features are extracted at multiple scales, varying their parameters. For the LBP descriptor, the selected parameters are $P = \{8, 16, 24\}$ and $R = \{1, 2, 3\}$. For LPQ and BSIF descriptors, the filter sizes were selected as $W =$

$\{3, 5, 7, 9, 11, 13, 15, 17\}$. *Deep learning features* Deep neural networks have been recently outperforming the state of the art in various classification tasks. Particularly, convolutional neural networks (CNNs) demonstrated impressive performance in object classification in general and face recognition in particular. However, deep neural networks require a huge amount of training data to learn efficient features. Unfortunately, this is not the case for the currently available kinship databases. We conducted preliminary experiments using a Siamese CNN architecture as well as a deep architecture proposed by a previous work [35]. As expected both approaches resulted in lower performance than using shallow features, due to the lack of enough training data.

An alternative for extracting deep face features is to use a pre-trained network. A number of very deep pre-trained architectures has already been made available to the research community. Motivated by the similarities between face recognition and kinship verification problems, where the goal is to compute the common features in two facial representations, we decided to use the VGG-face [27] network. VGG-face has been initially trained for face recognition on a reasonably large dataset of 2.6 million images of over 2622 people. This network has been evaluated for face verification from both pairs of images and videos showing interesting performance compared against state of the art.

The detailed parameters of the VGG-face CNN are provided by Table 4. The input of the network is an RGB face image of size 224×224 pixels. The network is composed of 13 linear convolution layers *(conv)*, each followed by a nonlinear rectification layer *(relu)*. Some of these rectification layers are followed by a nonlinear max pooling layer *(mpool)*. Following are two fully connected layers *(fc)* both outputting a vector of size 4096. At the top of the initial network are a *fully connected* layer with the size of classes to predict (2622) and a *softmax* layer for computing the class posterior probabilities.

In this context, to extract deep face features for kinship verification, we input the video frames one by one to the CNN and collect the feature vector issued by the fully connected layer fc7 (all the layers of the CNN except the class predictor fc8 layer and the softmax layer are used). Finally, all the frames' features of a given face video are averaged, resulting in a video descriptor that can be used for classification.

To classify a pair of face features as positive (the two persons have a kinship relation) or negative (no kinship relation between the two persons), we use a bi-class linear support vector machine classifier (SVM). Before feeding the features to the SVM, each pair of features has to be transformed into a single feature vector as imposed by the classifier. We have examined various ways for combining a pair of features, such as concatenation and vector distances. We have empirically found that utilizing the normalized absolute difference shows the best performance. Therefore, in our experiments, a pair of feature vectors $X = \{x_1, \ldots, x_d\}$ and $Y = \{y_1, \ldots, y_d\}$ is represented by the vector $F = \{f_1, \ldots, f_d\}$ where:

$$f_i = \sum_j \frac{|x_j - y_j|}{\sum_j (x_j + y_j)} \tag{9}$$

Table 4 VGG-face CNN architecture

Layer	0	1	2	3	4	5	6	7	8	9	10	11	12	13	14	15	16	17	18
Type	Input	conv	relu	conv	relu	mpool	conv	relu	conv	relu	mpool	conv	relu	conv	relu	conv	relu	mpool	conv
Name		conv1_1	relu1_1	conv1_2	relu1_2	pool1	conv2_1	relu2_1	conv2_2	relu2_2	pool2	conv3_1	relu3_1	conv3_2	relu3_2	conv3_3	relu3_3	pool3	conv4_1
Support		3	1	3	1	2	3	1	3	1	2	3	1	3	1	3	1	2	3
Filt dim		3		64			64		128			128		256		256			256
Num filts		64		64			128		128			256		256		256			512
Stride		1	1	1	1	2	1	1	1	1	2	1	1	1	1	1	1	2	1
Pad		1	0	1	0	0	1	0	1	0	0	1	0	1	0	1	0	0	1
Layer	19	20	21	22	23	24	25	26	27	28	29	30	31	32	33	34	35	36	37
Type	relu	conv	relu	conv	relu	mpool	conv	relu	conv	relu	conv	relu	mpool	conv	relu	conv	relu	conv	softmx
Name	relu4_1	conv4_2	relu4_2	conv4_3	relu4_3	pool4	conv5_1	relu5_1	conv5_2	relu5_2	conv5_3	relu5_3	pool5	fc6	relu6	fc7	relu7	fc8	prob
Support	1	3	1	3	1	2	3	1	3	1	3	1	2	7	1	1	1	1	1
Filt dim		512		512			512		512		512			512		4096		4096	
Num filts		512		512			512		512		512			4096		4096		2622	
Stride	1	1	1	1	1	2	1	1	1	1	1	1	2	1	1	1	1	1	1
Pad	0	1	0	1	0	0	1	0	1	0	1	0	0	0	0	0	0	0	0

4.2 Experimental Analysis

To evaluate the proposed approach, we use UvA-NEMO Smile database [10], which is currently the only available video kinship database. The database was initially collected for analyzing posed versus spontaneous smiles of subjects. Videos are recorded with a resolution of 1920 × 1080 pixels at a rate of 50 frames per second under controlled illumination conditions. A color chart is placed on the background of the videos to allow further illumination and color normalization. The videos are collected in controlled conditions and do not show any kind of bias [3]. The ages of the subjects in the database vary from 8 to 76 years. Many families participated in the database collection, allowing its use for evaluation of automatic kinship from videos. A total of 95 kin relations were identified between 152 subjects in the database. There are seven different kin relations between pairs of videos: sister-sister (S-S), brother-brother (B-B), sister-brother (S-B), mother-daughter (M-D), mother-son (M-S), father-daughter (F-D), and father-son (F-S). The association of the videos of persons having kinship relations gives 228 pairs of spontaneous and 287 pairs of posed smile videos. The statistics of the database are summarized in Table 5.

Following [11], we randomly generate negative kinship pairs corresponding to each positive pair. Therefore, for each positive pair, we associate the first video with a video of another person within the same kin subset while ensuring there is no relation between the two subjects. Examples of the positive pairs and the generated negative pairs are illustrated by Fig. 4. For all the experiments, we perform a per-relationship evaluation and report the average of spontaneous and posed videos. The accuracy for the whole database, by pooling all the relations, is also provided. Since the number of pairs of each relation is small, we apply a *leave-one-out* evaluation scheme.

We have performed various experiments to assess the performance of the proposed approach. In the following, we present and analyze the reported results.

Deep features against shallow features: First we compare the performance of deep features against the spatiotemporal features. The results for different features are reported in Table 6. The ROC curves for separate relations as well as for the

Table 5 Kinship statistics of UvA-NEMO Smile database

	Spontaneous		Posed	
Relation	Subj. #	Vid. #	Sub. #	Vid. #
S-S	7	22	9	32
B-B	7	15	6	13
S-B	12	32	10	34
M-D	16	57	20	76
M-S	12	36	14	46
F-D	9	28	9	30
F-S	12	38	19	56
All	75	228	87	287

Fig. 4 Samples of images form UvA-NEMO Smile database

Table 6 Accuracy (in %) of kinship verification using spatiotemporal and deep features on UvA-NEMO Smile database

Method	S-S	B-B	S-B	M-D	M-S	F-D	F-S	Mean	Whole set
BSIFTOP	75.07	83.46	71.23	82.46	72.37	81.67	79.84	78.01	75.83
LPQTOP	69.67	78.21	82.54	71.71	83.30	78.57	83.91	78.27	76.02
LBPTOP	80.47	77.31	70.50	78.29	72.37	84.40	71.50	76.41	72.82
DeepFeat	**88.92**	**92.82**	**88.47**	**90.24**	**85.69**	**89.70**	**92.69**	**89.79**	**88.16**

whole database are depicted in Fig. 5. The performances of the three spatiotemporal features (LBPTOP, LPQTOP, and BSIFTOP) show competitive results on different kinship relations. Considering the average accuracy and the accuracy of the whole set, LPQTOP is the best-performing method, closely followed by the BSIFTOP, while LBPTOP shows the worst performance.

On the other hand, deep features report the best performance on all kinship relations significantly improving the verification accuracy. The gain in verification performance of the deep features varies between 2% and 9%, for different relations, when compared to the best spatiotemporal accuracy. These results highlight the ability of CNNs in learning face descriptors. Even though the network has been trained for face recognition, the extracted face deep features are highly discriminative when used in the kinship verification task.

Comparing relations: The best verification accuracy is obtained for B-B and F-S, while the lowest are S-B and M-S. These results are maybe due to the different sexes of the pairs. One can conclude that checking the kinship relation is easier between persons of the same gender. However, a further analysis of this point is needed as the accuracy of S-S is average in our case. It is also remarkable that the performance of kinship between males (B-B and F-S) is better than between females (M-D and S-S). Moreover, large age differences between the persons composing a pair have an effect on the kinship verification accuracy. For instance, the age difference of brothers (best performance) is lower than it is for M-S (lowest performance).

Videos vs. images We have carried out an experiment to check if verifying kinship relations from videos instead of images is worthy. Therefore, we employ the first frame from each video of the database. For this experiment, spatial variants of

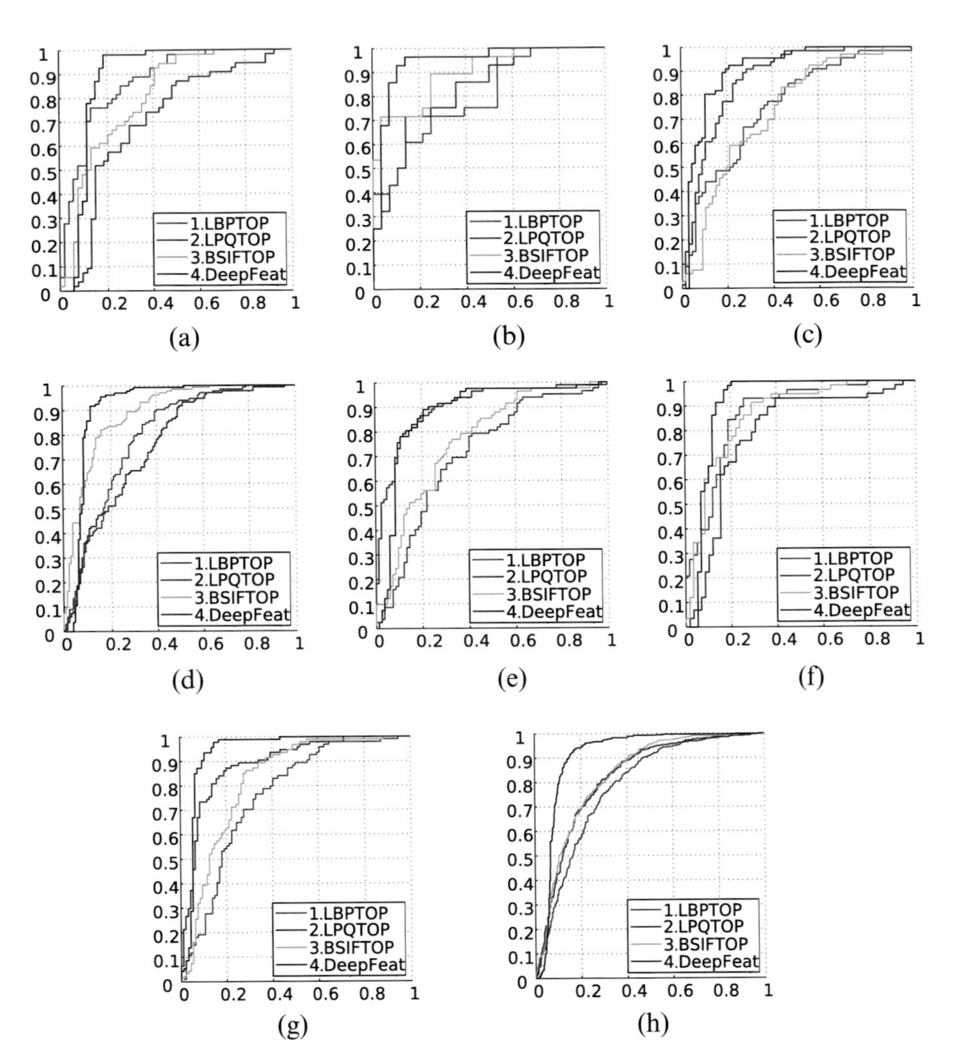

Fig. 5 Comparing deep vs. shallow features on UvA-NEMO Smile database. (**a**) Sister-Sister. (**b**) Brother-Brother. (**c**) Sister-Brother. (**d**) Mother-Daughter. (**e**) Mother-Son. (**f**) Father-Daughter. (**g**) Father-Son. (**h**) All

texture features (LBP, LPQ, and BSIF) and deep features are extracted from the face images. Figure 6 shows the ROC curve comparing the performance of videos against still images for the pool of all relationships. The superiority of the performance of videos compared with still images is obvious for each feature, demonstrating the importance of face dynamics in verifying kinship between persons. Again, deep features extracted from still face images demonstrate high discriminative ability, outperforming both the spatial texture features extracted from images and the spatiotemporal features extracted from videos. We note that, in still images (see

Fig. 6 Comparing videos vs. still images for kinship verification on UvA-NEMO Smile database

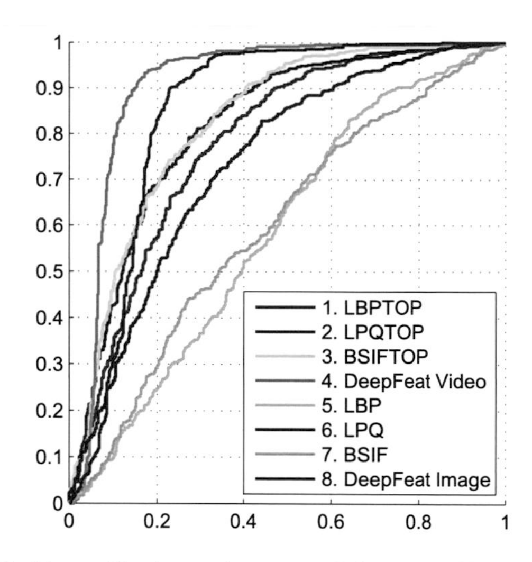

Table 7 Comparison of our approach for kinship verification against state of the art on UvA-NEMO Smile database

Method	S-S	B-B	S-B	M-D	M-S	F-D	F-S	Mean	Whole set
Fang et al. [12]	61.36	56.67	56.25	56.14	55.56	57.14	55.26	56.91	53.51
Guo and Wang [14]	65.91	56.67	60.94	58.77	62.50	67.86	55.26	61.13	56.14
Zhou et al. [39]	63.64	70.00	60.94	57.02	56.94	66.07	60.53	62.16	58.55
Dibeklioglu et al. [11]	75.00	70.00	68.75	67.54	75.00	75.00	78.95	72.89	67.11
Our DeepFeat	88.92	92.82	88.47	90.24	85.69	89.70	**92.69**	89.79	88.16
Our Deep + Shallow	**88.93**	**94.74**	**90.07**	**91.23**	**90.49**	**93.10**	88.30	**90.98**	**88.93**

Fig. 6), LPQ features outperform both LBP and BSIF, achieving analogous results to the ones computed using video data.

Feature fusion and comparison against state of the art: In order to check their complementarity, we have fused spatiotemporal features and deep features. We performed preliminary experiments and empirically found that score-level fusion performs better than feature fusion. In this context, and for simplicity, we have opted for a simple sum at the score level to perform the fusion. Table 7 shows a comparison of the fusion results with the previous works. Overall, the proposed fusion scheme improved further the verification accuracy by a significant margin. This effect is more evident in the relationships depicted by different sex and higher age variations, such as M-S (improved by 4.8%) and F-D (improved by 3.4%).

Comparing our results against the previously reported state of the art demonstrates considerable improvements in all the kinship subsets, as shown in Table 7. Depending on the relation type, the improvement in verification accuracy of our approach compared with the best-performing method presented by Dibeklioglu et al. [11] ranges from 9% to 23%. The average accuracy of all the kin relations has been improved by over 18%.

5 Conclusion

In this chapter, we have investigated the kinship verification problem from deep learning in both static and dynamic points of view, which demonstrates the high efficiency of deep features in describing faces for inferring kinship relations. A comparison of proposed approaches against the previous state-of-the-art work indicates significant improvements in verification accuracy.

It is worth noting that one can expect much better results with deep learning when a large number of training images are available for training. So, it is of great interest to re-evaluate the performance of our method when retraining our models on larger databases. Future work includes the collection of a large kinship database including real-world challenges to enable learning deep features.

References

1. Ahonen, T., Hadid, A., Pietikainen, M.: Face description with local binary patterns: Application to face recognition. IEEE Trans. Pattern Anal. Mach. Intell. **28**(12), 2037–2041 (2006)
2. Ahonen, T., Rahtu, E., Ojansivu, V., Heikkila, J.: Recognition of blurred faces using local phase quantization. In: Int. Conf. on Pattern Recognition, pp. 1–4 (2008)
3. Bordallo Lopez, M., Boutellaa, E., Hadid, A.: Comments on the "kinship face in the wild" data sets. IEEE Trans. Pattern Anal. Mach. Intell. (2016). https://doi.org/10.1109/TPAMI.2016.2522416
4. Bottino, A.G., Ul Islam, I., Vieira, T.: A multi-perspective holistic approach to kinship verification in the wild (2015)
5. Chan, C.H., Tahir, M., Kittler, J., Pietikainen, M.: Multiscale local phase quantization for robust component-based face recognition using kernel fusion of multiple descriptors. IEEE Trans. Pattern Anal. Mach. Intell. **35**(5), 1164–1177 (2013)
6. Chen, X., An, L., Yang, S., Wu, W.: Kinship verification in multi-linear coherent spaces. Multimedia Tools and Applications pp. 1–18 (2015)
7. Chopra, S., Hadsell, R., Lecun, Y.: Learning a similarity metric discriminatively, with application to face verification. In: Computer Vision and Pattern Recognition, 2005. CVPR 2005. IEEE Computer Society Conference on, pp. 539–546 vol. 1 (2005)
8. DalMartello, M.F., Maloney, L.T.: Where are kin recognition signals in the human face? Journal of Vision **6**(12), 2 (2006)
9. DeBruine, L.M., Smith, F.G., Jones, B.C., Roberts, S.C., Petrie, M., Spector, T.D.: Kin recognition signals in adult faces. Vision Research **49**(1), 38–43 (2009)
10. Dibeklioğlu, H., Salah, A., Gevers, T.: Are you really smiling at me? spontaneous versus posed enjoyment smiles. In: Computer Vision-ACCV, pp. 525–538. Springer (2012)
11. Dibeklioğlu, H., Salah, A., Gevers, T.: Like father like son facial expression dynamics for kinship verification. In: IEEE Int. Conf. on Computer Vision, pp. 1497–1504 (2013)
12. Fang, R., Tang, K., Snavely, N., Chen, T.: Towards computational models of kinship verification. In: IEEE Int. Conf. on Image Processing, pp. 1577–1580 (2010). https://doi.org/10.1109/ICIP.2010.5652590
13. Fang, R., Tang, K.D., Snavely, N., Chen, T.: Towards computational models of kinship verification. In: Proceedings / ICIP ... International Conference on Image Processing, pp. 1577–1580 (2010)
14. Guo, G., Wang, X.: Kinship measurement on salient facial features. IEEE Instrum. Meas. Mag. **61**(8), 2322–2325 (2012). https://doi.org/10.1109/TIM.2012.2187468

15. Hu, J., Lu, J., Yuan, J., Tan, Y.P.: Large margin multi-metric learning for face and kinship verification in the wild. In: Computer Vision–ACCV 2014, pp. 252–267. Springer (2015)
16. Huang, G., Lee, H., Learned-Miller, E.: Learning hierarchical representations for face verification with convolutional deep belief networks. In: Computer Vision and Pattern Recognition (CVPR), 2012 IEEE Conference on, pp. 2518–2525 (2012). https://doi.org/10.1109/CVPR.2012.6247968
17. Kannala, J., Rahtu, E.: BSIF: Binarized statistical image features. In: Int. Conf. on Pattern Recognition (ICPR), pp. 1363–1366 (2012)
18. Kou, L., Zhou, X., Xu, M., Shang, Y.: Learning a genetic measure for kinship verification using facial images. Mathematical Problems in Engineering **2015** (2015)
19. Lu, J., Zhou, X., Tan, Y.P., Shang, Y., Zhou, J.: Neighborhood repulsed metric learning for kinship verification. In: Computer Vision and Pattern Recognition (CVPR), 2012 IEEE Conference on, pp. 2594–2601 (2012)
20. Lu, J., Zhou, X., Tan, Y.P., Shang, Y., Zhou, J.: Neighborhood repulsed metric learning for kinship verification. Pattern Analysis and Machine Intelligence, IEEE Transactions on **36**(2), 331–345 (2014)
21. Lu, J., et al.: Kinship verification in the wild: The first kinship verification competition. In: IEEE Int. Joint Conf. on Biometrics (IJCB), pp. 1–6 (2014)
22. Lu, J., Hu, J., Zhou, X., Zhou, J., Castrillón-Santana, M., Lorenzo-Navarro, J., Kou, L., Shang, Y., Bottino, A., Figuieiredo Vieira, T.: Kinship verification in the wild: The first kinship verification competition. In: Biometrics (IJCB), 2014 IEEE International Joint Conference on, pp. 1–6. IEEE (2014)
23. Lu, J., Hu, J., Liong, V.E., Zhou, X., Bottino, A., Islam, I.U., Vieira, T.F., Qin, X., Tan, X., Chen, S., et al.: The fg 2015 kinship verification in the wild evaluation. Proc. FG (2015)
24. Lu, J., et al.: The FG 2015 kinship verification in the wild evaluation. In: IEEE Int. Conf. on Automatic Face and Gesture Recognition (FG), vol. 1, pp. 1–7 (2015). https://doi.org/10.1109/FG.2015.7163159
25. Mignon, A.: Pcca: A new approach for distance learning from sparse pairwise constraints. In: 2012 IEEE Conference on Computer Vision and Pattern Recognition, pp. 2666–2672 (2012)
26. Palm, R.B.: Prediction as a candidate for learning deep hierarchical models of data (2012). http://www.imm.dtu.dk/English.aspx. Supervised by Associate Professor Ole Winther, owi@imm.dtu.dk, DTU Informatics, and Morten Mørup, mm@imm.dtu.dk, DTU Informatics
27. Parkhi, O.M., Vedaldi, A., Zisserman, A.: Deep face recognition. In: British Machine Vision Conf. (2015)
28. Qin, X., Tan, X., Chen, S.: Tri-subject kinship verification: understanding the core of a family. Multimedia, IEEE Transactions on **17**(10), 1855–1867 (2015)
29. Somanath, G., Kambhamettu, C.: Can faces verify blood-relations? In: Biometrics: Theory, Applications and Systems (BTAS), 2012 IEEE Fifth International Conference on, pp. 105–112. IEEE (2012)
30. Wang, X., Kambhamettu, C.: Leveraging appearance and geometry for kinship verification. In: Image Processing (ICIP), 2014 IEEE International Conference on, pp. 5017–5021. IEEE (2014)
31. Xia, S., Shao, M., Fu, Y.: Kinship verification through transfer learning. In: IJCAI Proceedings-International Joint Conference on Artificial Intelligence, vol. 22, p. 2539 (2011)
32. Xia, S., Shao, M., Fu, Y.: Toward kinship verification using visual attributes. In: Pattern Recognition (ICPR), 2012 21st International Conference on, pp. 549–552. IEEE (2012)
33. Yan, H., Lu, J., Deng, W., Zhou, X.: Discriminative multimetric learning for kinship verification. Information Forensics and Security, IEEE Transactions on **9**(7), 1169–1178 (2014)
34. Yan, H.C., Lu, J., Zhou, X.: Prototype-based discriminative feature learning for kinship verification (2014)
35. Zhang, K., Huang, Y., Song, C., Wu, H., Wang, L.: Kinship verification with deep convolutional neural networks. In: British Machine Vision Conf. (BMVC) (2015)
36. Zhao, G., Pietikainen, M.: Dynamic texture recognition using local binary patterns with an application to facial expressions. IEEE Trans. Pattern Anal. Mach. Intell. **29**(6), 915–928 (2007)

37. Zhou, X., Hu, J., Lu, J., Shang, Y., Guan, Y.: Kinship verification from facial images under uncontrolled conditions. In: ACM Int. Conf. on Multimedia, pp. 953–956 (2011)
38. Zhou, X., Lu, J., Hu, J., Shang, Y.: Gabor-based gradient orientation pyramid for kinship verification under uncontrolled environments. In: Proceedings of the 20th ACM international conference on Multimedia, pp. 725–728. ACM (2012)
39. Zhou, X., Lu, J., Hu, J., Shang, Y.: Gabor-based gradient orientation pyramid for kinship verification under uncontrolled environments. In: ACM Int. Conf. on Multimedia, pp. 725–728 (2012). https://doi.org/10.1145/2393347.2396297.
40. Zhou, X., Shang, Y., Yan, H., Guo, G.: Ensemble similarity learning for kinship verification from facial images in the wild. Information Fusion (2015). https://doi.org/10.1016/j.inffus.2015.08.006. http://www.sciencedirect.com/science/article/pii/S1566253515000792

Deep Learning Architectures for Face Recognition in Video Surveillance

Saman Bashbaghi, Eric Granger, Robert Sabourin, and Mostafa Parchami

Abstract Face recognition (FR) systems for video surveillance (VS) applications attempt to accurately detect the presence of target individuals over a distributed network of cameras. In video-based FR systems, facial models of target individuals are designed a priori during enrollment using a limited number of reference still images or video data. These facial models are not typically representative of faces being observed during operations due to large variations in illumination, pose, scale, occlusion, blur, and camera interoperability. Specifically, in still-to-video FR application, a single high-quality reference still image captured with still camera under controlled conditions is employed to generate a facial model to be matched later against lower-quality faces captured with video cameras under uncontrolled conditions. Current video-based FR systems can perform well on controlled scenarios, while their performance is not satisfactory in uncontrolled scenarios mainly because of the differences between the source (enrollment) and the target (operational) domains. Most of the efforts in this area have been toward the design of robust video-based FR systems in unconstrained surveillance environments. This chapter presents an overview of recent advances in still-to-video FR scenario through deep convolutional neural networks (CNNs). In particular, deep learning architectures proposed in the literature based on triplet-loss function (e.g., cross-correlation matching CNN, trunk-branch ensemble CNN and HaarNet) and supervised autoencoders (e.g., canonical face representation CNN) are reviewed and compared in terms of accuracy and computational complexity.

S. Bashbaghi · E. Granger (✉) · R. Sabourin
Laboratoire. d'imagerie de vision et d'intelligence artificielle, École de technologie supérieure, Université du Québec, Montreal, QC, Canada
e-mail: bashbaghi@livia.etsmtl.ca; eric.granger@etsmtl.ca; robert.sabourin@etsmtl.ca

M. Parchami
Computer Science and Engineering Department, University of Texas at Arlington, Arlington, TX, USA
e-mail: mostafa.parchami@mavs.uta.edu

X. Jiang et al. (eds.), *Deep Learning in Object Detection and Recognition*,
https://doi.org/10.1007/978-981-10-5152-4_6

1 Introduction

Face recognition (FR) systems in video surveillance (VS) have received a significant attention during the past few years. Due to the fact that the number of surveillance cameras installed in public places is increasing, it is important to build robust video-based FR systems [38]. In VS, capture conditions typically range from semi-controlled with one person in the scene (e.g., passport inspection lanes and portals at airports) to uncontrolled free-flow in cluttered scenes (e.g., airport baggage claim areas, and subway stations). Two common types of applications in VS are (1) still-to-video FR (e.g., watch-list screening) and (2) video-to-video FR (e.g., face re-identification or search and retrieval) [4, 11, 23]. In the former application, reference face images or stills of target individuals of interest are used to design facial models, while in the latter, facial models are designed using faces captured in reference videos. This chapter is mainly focused on still-to-video FR with a single sample per person (SSPP) under semi- and unconstrained VS environments.

The number of target references is one or very few in still-to-video FR applications, and the characteristics of the still camera(s) used for design significantly differ from the video cameras used during operations [3]. Thus, there are significant differences between the appearances of still ROI(s) and ROIs captured with surveillance cameras, according to various changes in ambient lighting, pose, blur, and occlusion [1, 21]. During enrollment of target individuals, facial regions of interests (ROIs) isolated in reference still images are used to design facial models, while during operations, the ROIs of faces captured in videos are matched against these facial models. In VS, a person in a scene may be tracked along several frames, and matching scores may be accumulated over a facial trajectory (a group of ROIs that correspond to the same high-quality track of an individual) for robust spatiotemporal FR [7].

In general, methods proposed in the literature for still-to-video FR can be broadly categorized into two main streams: (1) conventional and (2) deep learning methods. The conventional methods rely on hand-crafted feature extraction techniques and a pre-trained classifier along with fusion, while deep learning methods automatically learn features and classifiers cojointly using massive amounts of data. In spite of improvements achieved using the conventional methods, yet they are less robust to real-world still-to-video FR scenario. On the other hand, there exists no feature extraction technique that can overcome all the challenges encountered in VS individually [4, 15, 34].

Conventional methods proposed for still-to-video FR are typically modeled as individual-specific face detectors using one- or two-class classifiers in order to enable the system to add or remove other individuals and easily adapt over time [2, 23]. Modular systems designed using individual-specific ensembles have been successfully applied in VS [11, 23]. Thus, ensemble-based methods have been shown as a reliable solution to deal with imbalanced data, where multiple face representations can be encoded into ensembles of classifiers to improve the robustness of still-to-video FR [4]. Although it is challenging to design robust facial models using a single training sample, several approaches have addressed this problem, such as multiple face representations, synthetic generation of virtual faces, and using auxiliary data from other people to enlarge the training set [2, 16, 17, 36]. These techniques seek to enhance the robustness of face models to intra-class variations. In multiple representations, different patches and face descriptors are employed [2, 4], while 2D morphing or 3D reconstructions are used to synthesize artificial face images [16, 22]. A generic auxiliary dataset containing faces of other persons can be exploited to perform domain adaptation [20] and sparse representation classification through dictionary learning [36]. However, techniques based on synthetic face generation and auxiliary data are more complex and computationally costly for real-time applications, because of the prior knowledge required to locate the facial components reliably, and the large differences between the quality of still and video ROIs, respectively.

Recently, several deep learning-based solutions have been proposed to learn effective face representations directly from training data through convolutional neural networks (CNNs) and nonlinear feature mappings [6, 14, 28, 29, 31]. In such methods, different loss functions can be considered in the training process to enhance the interpersonal variations and simultaneously reduce the intrapersonal variations. They can learn nonlinear and discriminative feature representations to cover the existing gaps compared to the human visual system [34], while they are computationally costly and typically require a large number of labeled data to train. To address the SSPP problem in FR, a triplet-based loss function has been introduced in [8, 24, 25, 27, 28] to discriminate between a pair of matching ROIs and a pair of nonmatching ROIs. Ensemble of CNNs, such as trunk-branch ensemble CNN (TBE-CNN) [8] and HaarNet [25], has been shown to extract features from the global appearance of faces (holistic representation), as well as to embed asymmetrical features (local facial feature-based representations) to handle partial occlusion. Moreover, supervised autoencoders have been proposed to enforce faces with variations to be mapped to the canonical face (a well-illuminated frontal face with neutral expression) of the person in the SSPP scenario to generate robust feature representations [9, 26].

2 Background of Video-Based FR Through Deep Learning

Deep CNNs have recently demonstrated a great achievement in many computer vision tasks, such as object detection, object recognition, etc. Such deep CNN models have shown to appropriately characterize different variations within a large amount of data and to learn a discriminative nonlinear feature representation. Furthermore, they can be easily generalized to other vision tasks by adopting and fine-tuning pre-trained models through transfer learning [6, 28]. Thus, they provide a successful tool for different applications of FR by learning effective feature representations directly from the face images [6, 14, 28]. For example, DeepID,

DeepID2, and DeepID2+ have been proposed in [30–32], respectively, to learn a set of discriminative high-level feature representations.

For instance, an ensemble of CNN models was trained in [31] using the holistic face image along with several overlapping/nonoverlapping face patches to handle the pose and partial occlusion variations. Fusion of these models is typically carried out by feature concatenation to construct over-complete and compact representations. Followed by [31], feature dimension of the last hidden layer was increased in [30, 32], as well as exploiting supervision to the convolutional layers in order to learn hierarchical and nonlinear feature representations. These representations aim to enhance the interpersonal variations due to extraction of features from different identities separately and simultaneously reduce the intrapersonal variations. In contrast to DeepID series, an accurate face alignment was incorporated in Microsoft DeepFace [34] to derive a robust face representation through a nine-layer deep CNN. In [29], the high-level face similarity features were extracted jointly from a pair of faces instead of a single face through multiple deep CNNs for face verification applications. Since these approaches are not considered variations like blurriness and scale changes (distance of the person from surveillance cameras), they are not fully adapted for video-based FR applications.

Similarly, for the SSPP problems, a triplet-based loss function has been lately exploited in [8, 24, 25, 27, 28] to learn robust face embeddings, where this type of loss seeks to discriminate between the positive pair of matching facial ROIs and the negative nonmatching facial ROI. A robust facial representation learned through triplet-loss optimization has been proposed in [24] using a compact and fast cross-correlation matching CNN (CCM-CNN). However, CNN models like the trunk-branch ensemble CNN (TBE-CNN) [8] and HaarNet [25] can further improve robustness to variations in facial appearance by the cost of increasing computational complexity. In such models, the trunk network extracts features from the global appearance of faces (holistic representation), while the branch networks embed asymmetrical and complex facial traits. For instance, HaarNet employs three branch networks based on Haar-like features, while facial landmarks are considered in TBE-CNN. However, these specialized CNNs represent complex solutions that are not perfectly suitable for real-time FR applications [5].

Moreover, autoencoder neural networks can be typically employed to extract deterministic nonlinear feature mappings robust to face images contaminated by different noises, such as illumination, expression, and poses [9, 26]. An autoencoder network contains encoder and decoder modules, where the former module embeds the input data to the hidden nodes, while the latter returns the hidden nodes to the original input data space with minimizing the reconstruction error(s) [9]. Several autoencoder networks inspired from [35] have been proposed to remove the aforementioned variances in face images [9, 18, 19]. These networks deal with faces containing different types of variations (e.g., illumination, pose, etc.) as noisy images. For instance, a facial component-based CNN has been learned in [40] to transform faces with changes in pose and illumination to frontal view faces, where pose-invariant features of the last hidden layer are employed as face representations. Similarly, several deep architectures have been proposed using multitask learning in order to rotate faces with arbitrary poses and illuminations to target-pose faces [37, 39]. In addition, a general deep architecture was introduced in [10] to encode a desired attribute and combine it with the input image to generate target images as similar as the input image with a visual attribute (a different illumination, facial appearance, or new pose) without changing other aspects of a face.

3 Deep Learning Architectures for FR in VS

In this section, the most recent deep learning architectures proposed for video-based FR considering the SSPP problem are addressed. These architectures can be categorized into two groups: (1) deep CNN models trained using triplet-loss function and (2) deep autoencoders.

3.1 Deep CNNs Using Triplet-Loss

Recently, deep learning algorithms specialized for FR mostly utilize triplet-loss in order to train the deep architecture and thereby learn a discriminant face representation [8, 28]. However, careful triplet sampling is a crucial step to achieve a faster convergence [28]. In addition, employing triplet-loss is challenging since the global distributions of the training samples are neglected in optimization process.

Triplet-loss approach was first proposed in [28] to train CNNs for robust face verification. To that end, the representation of triplets (three faces containing an anchor and a positive image of the same subject and a negative image of other subjects) is L_2 normalized as the input of triplet-loss function. It therefore ensures that the input representations of face images lie on a unit hypersphere prior to apply triplet-loss function [8]. Deep CNN models proposed for video-based FR that employed triplet-loss for training are reviewed in the following subsections.

3.1.1 Cross-Correlation Matching CNN

An efficient deep CNN architecture has been proposed in [24] for still-to-video FR from a single reference facial ROI per target individual. Based on a pair-wise cross-correlation matching (CCM) along with a robust facial representation learned through triplet-loss optimization, CCM-CNN is a fast and compact network (requires few branches, layers, and parameters). It exploits a matrix Hadamard product followed by a fully connected layer that simulates the adaptive weighted cross-correlation [12]. A triplet-based optimization approach has been employed to learn discriminant representations based on triplets containing the positive and negative video ROIs and the corresponding still ROI. In particular, the similarity between the representations of positive video ROIs and the reference still ROI is

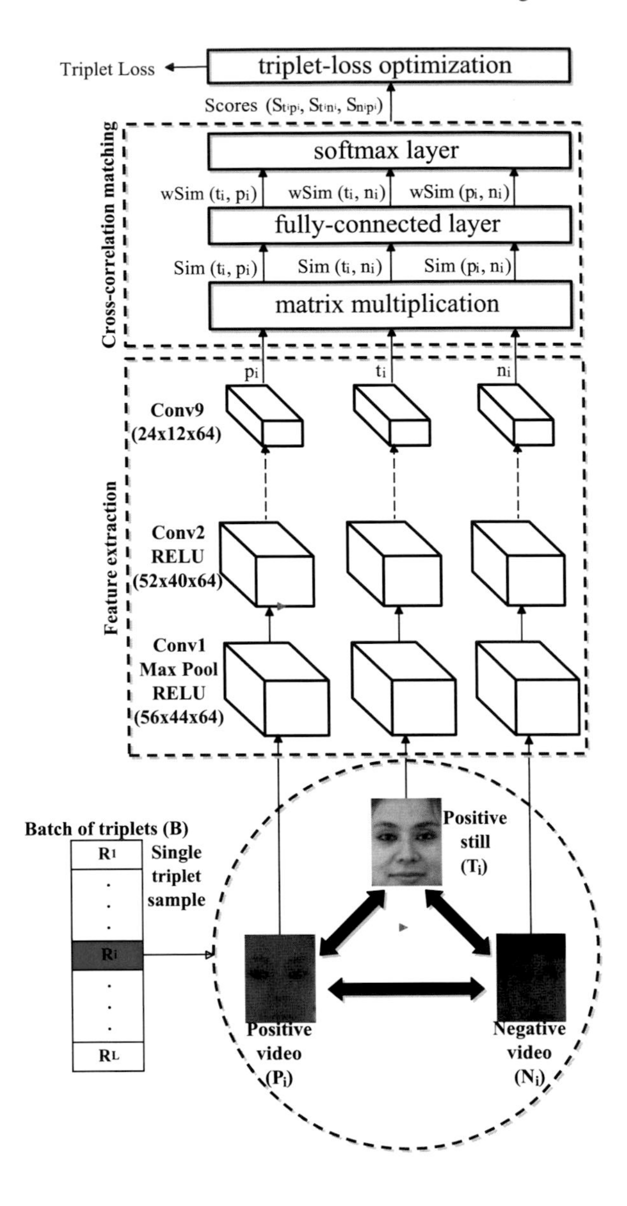

Fig. 1 Training pipeline of the CCM-CNN. (©[2017] IEEE. Reprinted, with permission, from Ref. [24])

enhanced, while the similarity between negative video ROIs and the both reference still and positive video ROIs is increased. To further improve robustness of facial models, the CCM-CNN fine-tuning process incorporates a diverse knowledge by generating synthetic faces based on still and video ROIs of nontarget individuals.

As shown in Fig. 1, the CCM-CNN learns a robust facial representation by iterating over a batch of training triplets $B = \{R_1, \ldots, R_L\} = \{(T_1, P_1, N_1), \ldots, (T_L, P_L, N_L)\}$, where L is the batch size, and each triplet R_i contains a still ROI T_i along with a corresponding positive ROI P_i and a negative ROI N_i from operational videos. This architecture was inspired by Siamese networks containing identical subnetworks with the same configurations, parameters, and weights. Therefore, fewer parameters are required for training that can avoid overfitting. The CCM-CNN consists of three main components – feature extraction, cross-correlation matching, and triplet-loss optimization. The feature extraction pipeline extracts discriminative feature maps from ROIs that are similar for two images of the same person under different capture conditions (e.g., illumination and pose). The cross-correlation matching component inputs feature maps extracted from the ROIs and calculates the likelihood of the faces belonging to the same person. Finally, triplet-loss optimization computes a loss function to maximize similarity of the still ROIs and their respective positive samples in the batch while minimizing similarity between still ROIs and their negative ROIs and positive and negative ROIs.

Despite differences in the domains between reference target still ROIs and target/nontarget video ROIs, the CCM-CNN can effectively extract discriminant features. As shown in Fig. 1, feature extraction is carried out by three identical subnetworks for still, positive, and negative faces. These subnetworks process three input faces, and the weights are shared across them. Each subnetwork consists of nine convolutional layers each followed by a spatial batch normalization, dropout, and RELU layers. Contrary to former convolutional layers, the last convolutional layer is not followed by a RELU in order to maintain the representativeness of the final feature map and to avoid losing informative data for the matching stage. Moreover, a single max-pooling layer is added after the first convolution layer to increase the robustness to small translation of faces in the ROI.

In the CCM-CNN, all three feature extraction pipelines share the same set of parameters. This ensures that the features extracted from target still ($\mathbf{t_i}$), positive ($\mathbf{p_i}$), and negative ($\mathbf{n_i}$) are consistent and comparable. Each convolutional layer has 64 filters of size 5×5 without padding. Thus, given the input size of 120×96, the output of each branch is of size $N_f = 24 \times 12 \times 64$ features.

After extracting features from the still and video ROIs, a pixel-based matching method is employed to effectively compare these feature maps and measure the matching similarity. The process of comparison in the CCM-CNN has three stages: matrix Hadamard product, fully connected neural network, and finally a softmax. Instead of concatenating feature vectors of different branches as input to the fully connected layer, the feature maps representing the ROIs are multiplied with each other to encode pixel-wise correlation between each pair of ROI in the triplet. This approach eliminates the complexity of matching by replacing the concatenation with a simple element-wise matrix multiplication and directly encodes similarity as opposed to let the network learn how to match input concatenated feature vectors.

The matrix Hadamard product is exploited to simulate cross-correlation, where Hadamard product of the two matrices provides a single feature map that represents the similarity of the two ROIs. For example, the similarity $Sim(\mathbf{t}_i, \mathbf{p}_i)$ and cross-correlation $\mathbf{w}Sim(\mathbf{t}_i, \mathbf{p}_i)$ of still $\mathbf{t}_i$ and positive $\mathbf{p}_i$ feature maps is computed as follows, respectively, using matrix Hadamard product:

$$Sim(\mathbf{t}_i, \mathbf{p}_i) = (\mathbf{t}_i \odot \mathbf{p}_j) \quad (1)$$

$$wSim(\mathbf{t}_i, \mathbf{p}_i) = \boldsymbol{\omega}_m \cdot RELU(\boldsymbol{\omega}_n \cdot Sim(\mathbf{t}_i, \mathbf{p}_i) + \mathbf{b}_n) + \mathbf{b}_m \quad (2)$$

where $\boldsymbol{\omega}_m$, $\boldsymbol{\omega}_n$, $\mathbf{b}_m$, and $\mathbf{b}_n$ are the weights and biases of the two fully connected layers applied to the vectorized output of the matrix multiplication. Furthermore, a softmax layer is applied to obtain a probability-like similarity score for each of the two classes (match and nonmatch).

A multistage approach is considered to efficiently train the CCM-CNN based on reference still ROI and operational videos. To that end, pre-training is performed using a large generic FR dataset, and a domain-specific dataset for still-to-video FR is used for fine-tuning. To that end, a set of matching and nonmatching images is selected from the Labeled Faces in the Wild (LFW) [13]. Images from this set are augmented to roughly 1.3M training triplets. In order to consistently update the set of training triplets, the online triplet sampling method [28] is used for 50 epochs.

In contrast with FaceNet [28], a pairwise triplet-loss optimization function was proposed to effectively train the network. In order to adapt the network for pairwise triplet-based optimization, it is modified by incorporating additional feature extraction branches. Each batch contains several triplets, and for each triplet, the network seeks to learn the correct classification. During the training, each branch of the feature extraction pipeline is assigned to a component of the triplet – the main branch is responsible for processing the reference still ROI, while the positive (negative) branch extracts features from the positive (negative) video ROI of the triplet. Moreover, the cross-correlation matching pipeline is modified to benefit from the triplets by introducing an Euclidean loss layer followed by softmax which computes the similarity for each pair of ROIs in the triplet. The loss layer is exploited to compute the overall loss of the network as follows:

$$\text{Triplet Loss} = \frac{1}{L} \sum_{R_i \in B} \sqrt{\left(1 - S_{t_i p_i}\right)^2 + S_{t_i n_i}^2 + S_{n_i p_i}^2} \quad (3)$$

where S_{tp}, S_{tn}, and S_{np} are the similarity scores from cross-correlation matching between (1) the reference (positive) still ROI and positive video ROI, (2) still ROI and negative video ROI, and (3) negative and positive video ROIs of the triplet, respectively, computed using the aforementioned approach. During operations (see Fig. 2) the additional feature extraction branch (negative branch, N) is removed from the network, and only the still and the positive branches (P) are taken into account. Thus, the main branch (T) extracts features from reference still ROIs, while the positive branch extracts features from the probe video ROI to determine whether they belong to the same person.

During fine-tuning, CCM-CNN acquires knowledge on the similarities and dissimilarities between the target individuals of interest enrolled to the system. In order to improve the robustness of facial models intra-class variation, the network is fine-tuned with synthetic facial ROIs generated from the high-quality still ROIs that

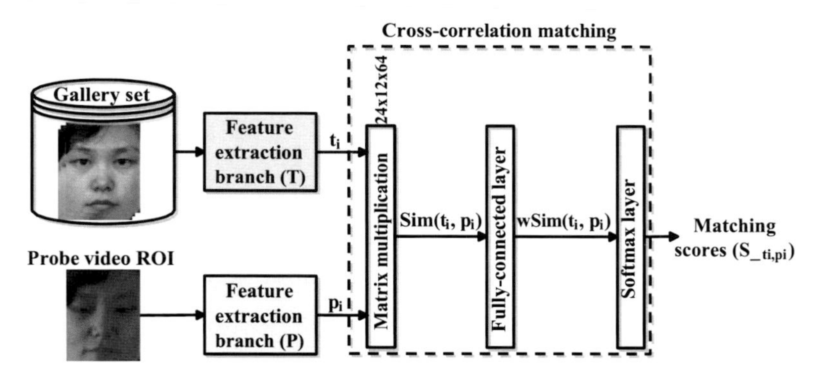

Fig. 2 The operational phase of the CCM-CNN

account for the operation domain. For each still image, a set of augmented images are generated using different transformations, such as shearing, mirroring, rotating, and translating the original still image. In contrast with the pre-training, the focus of the fine-tuning stage is to learn dissimilarities between the subjects of interest.

3.1.2 Trunk-Branch Ensemble CNN

An improved triplet-loss function has been introduced in [8] to promote the robustness of face representations. To that end, a trunk-branch ensemble CNN (TBE-CNN) model has been proposed to extract complementary features from holistic face images as well as face patches around facial landmarks through trunk and branch networks, respectively. To emulate real-world video data, artificially blur training data are synthesized from still images by applying artificial out-of-focus and motion blur to learn blur-insensitive face representations. The architecture of TBE-CNN is shown in Fig. 3.

As shown in Fig. 3, TBE-CNN contains one trunk network along with several branch networks, where the trunk and branch networks share some layers in order to embed global and local information. This sharing strategy may lead to reduce the computational cost and also efficient convergence. The output feature maps of these networks are concatenated to feed into the fully connected layer to generate final face representations.

During training as illustrated in Fig. 4, TBE-CNN is given still images and simulated video frames, where the network aims to classify each still image and its corresponding artificially blurred face image correctly into the same class. The training process is performed using a stage-wise strategy, where the trunk network and each of the branch networks are trained separately with fixed parameters.

To improve the discriminative power of face representations, mean distance regularized triplet-loss (MDR-TL) function is considered to fine-tune the entire network. Compared to the original triplet-loss function proposed in [28], MDR-

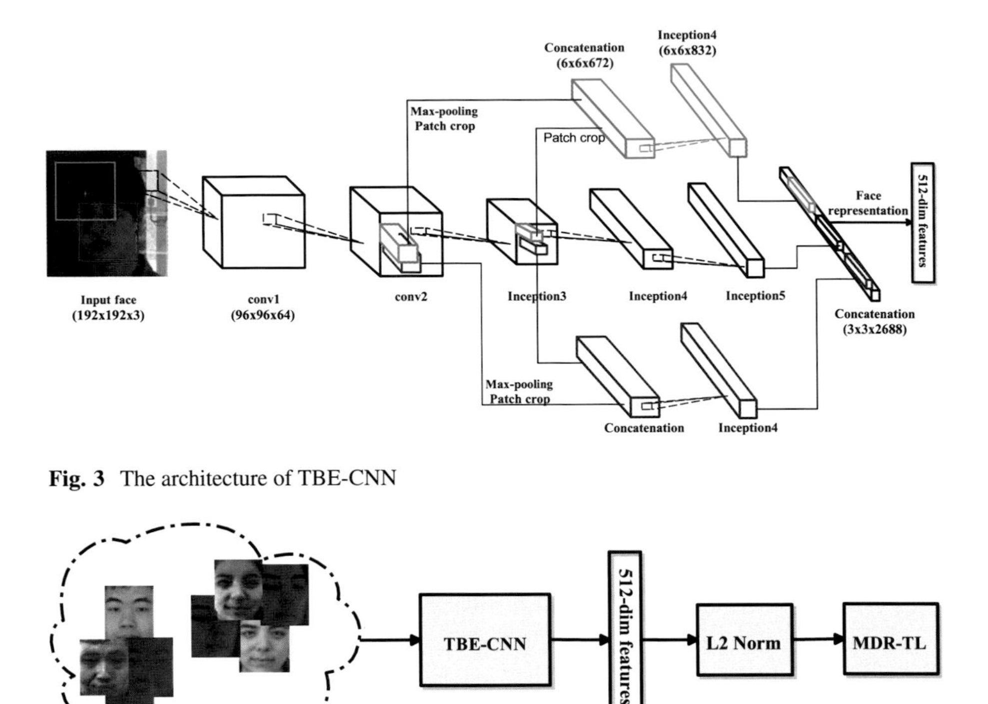

Fig. 3 The architecture of TBE-CNN

Fig. 4 Training pipeline of the TBE-CNN

TL regularizes the triplet-loss to provide uniform distributions for both inter- and intra-class distances. Figure 5 represents the principle of MDR-TL.

As demonstrated in Fig. 5a, it is difficult to appropriately discriminate between matching and nonmatching pairs of face images because the training samples have nonuniform inter- and intra-class distance distributions. To tackle this problem, the triplet-loss is regularized using MDR-TL loss function by constraining the distances between mean representations of different subjects (Fig. 5b).

3.1.3 HaarNet

An ensemble of deep CNNs called HaarNet has been proposed in [25] to efficiently learn robust and discriminative face representations for video-based FR applications. Similar to TBE-CNN [8], HaarNet consists of a trunk network with three diverging branch networks that are specifically designed to embed facial features, pose, and other distinctive features. The trunk network effectively learns a holistic representation of the face, whereas the branches learn more local and asymmetrical features related to pose or special facial features by means of Haar-like features.

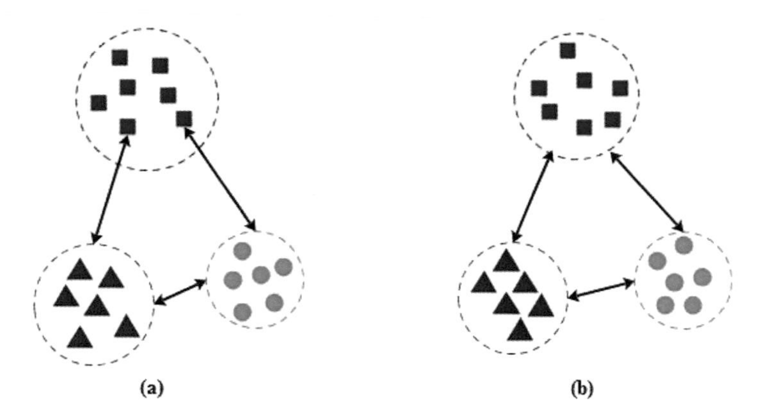

Fig. 5 The mean distance regularized triplet-loss. (**a**) Training triplet with nonuniform inter- and intra-class distance distributions and (**b**) triplets with uniform inter- and intra-class distance distributions using MDR-TL regularization

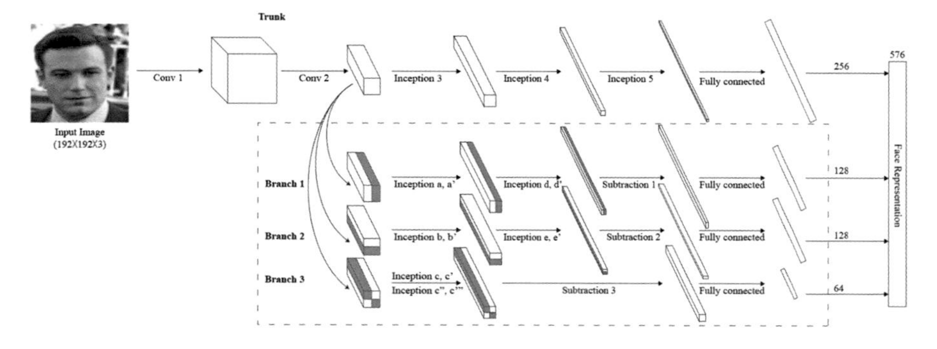

Fig. 6 HaarNet architecture for the trunk and three branches. (Max pooling layers after each inception and convolution layer are not shown for clarity). (©[2017] IEEE. Reprinted,with permission, from Ref.[25])

Furthermore, to increase the discriminative capabilities of the HaarNet, a second-order statistic regularized triplet-loss function has been introduced to take advantage of the inter-class and intra-class variations existing in training data to learn more distinctive representations for subjects with similar faces. Finally, a fine-tuning stage has been performed to embed the correlation of facial ROIs stored during enrollment and improve recognition accuracy.

The overall architecture of the HaarNet is presented in Fig. 6. It is composed of a global trunk network along with three branch networks that can effectively learn a representation that is robust to changing capture conditions.

As shown in Fig. 6, the trunk is employed to learn the global appearance face representation, whereas three branches diverged from the trunk are designed to learn asymmetrical and more locally distinctive representations. For the trunk network, the configuration of GoogLeNet [33] is employed with 18 layers.

In contrast with [8], instead of training each branch on different face landmarks, Haarnet utilizes three branch networks in order to compute one of the Haar-like features, respectively, as illustrated in Fig. 7. Haar features have been exploited to extract distinctive features from faces based on the symmetrical nature of facial components and on contrast of intensity between adjacent components. In general, these features are calculated by subtracting sum of all pixels in the black areas from the sum of all pixels in the white areas. To avoid information loss, the Haar-like features are calculated by matrix summation, where black matrices are negated. Thus, instead of generating only one value, each Haar-like feature returns a matrix.

In the Haarnet architecture (see Fig. 6), the trunk network and its three branches share the first two convolutional layers. Then, the first and second branches split the output of Conv2 into two sub-branches and also apply two inception layers to each sub-branch. Subsequently, the two sub-branches are merged by a subtraction layer to obtain a Haar-like representation for each corresponding branch. Meanwhile, the third branch divides the output of Conv2 into four sub-branches, and one inception layer is applied to each of the sub-branches. Eventually, a subtraction layer is exploited to combine those for sub-branches and feed to the fully connected layer. The final representation of the face is obtained by concatenating the output of the trunk and all three Haar-like features.

Figure 8 illustrates the training process of the HaarNet using a triplet-loss concept, where a batch of triplets composed of <anchor, positive, negative> is input to the architecture translated to a face representation.

As shown in Fig. 8, output of the HaarNet is then $L2$ normalized prior to feed into the triplet-loss function in order to represent faces on a unit hypersphere. Let's

Fig. 7 Haar-like features used in branch networks. (©[2017] IEEE. Reprinted, with permission, from Ref. [25])

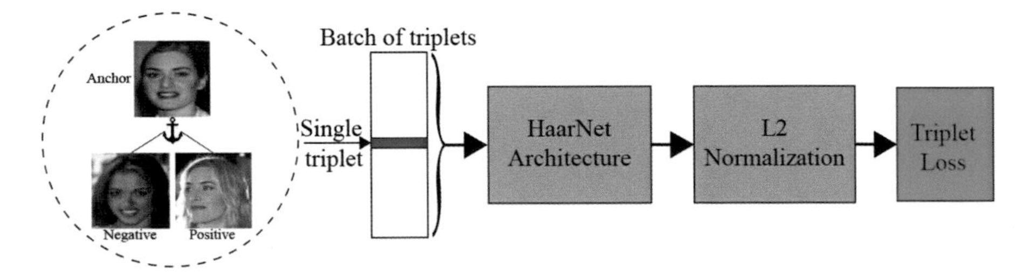

Fig. 8 Processing of triplets to compute the loss function. The network inputs a batch of triplets to the HaarNet architecture followed by an $L2$ Normalization. (©[2017] IEEE. Reprinted, with permission, from Ref. [25])

denote the $L2$ normalized representation of a facial ROI x as $f(x) \in R^d$ where d is the dimension of the face representation.

A multistage training approach is hereby considered to effectively optimize the parameters of the HaarNet. The first three stages are designed for initializing the parameters with a promising approximation prior to employ the triplet-loss function. Moreover, these three stages are beneficial to detect a set of hard triplets from the dataset in order to initiate the triplet-loss training. In the first stage, the trunk network is trained using a softmax loss, because the softmax function converges much faster than triplet-loss function. During the second stage, each branch is trained separately by fixing the shared parameters and by only optimizing the rest of the parameters. Similar to the first stage, a softmax loss function is used to train each of the branches. Then, the complete network is constructed by assembling the trunk and the three branch networks. The third stage of the training is indeed a fine-tuning stage for the complete network in order to optimize these four components simultaneously. In order to consider the inter- and intra-class variations, the network is trained for several epochs using the hard triplets detected during the previous stages.

As suggested in [8], adding mean distance regularization term to the triplet-loss function can promote distinctiveness of the face representations. Inspired from [8], the main idea of the second-order statistics regularization term is illustrated in Fig. 9 illustrates.

In Fig. 9a, triplet-loss function may suffer from nonuniform inter-class distances that leads to failure of using simple distance measures, such as Euclidean and cosine distances. In this regard (see Fig. 9b), a mean distance regularization term can be added to increase the separation of class representations. On the other hand, representations of some facial ROIs may be confused with representation of the adjacent facial ROIs in the feature space due to high intra-class variations. Figure 9c shows such a configuration, where the mean representation of the classes are distant from each other but the standard deviations of classes are very high, leading to overlap among class representations. To address this issue, a new term in the loss function is introduced to examine the intra-class distribution of the training samples.

The triplet constraint can be expressed as a function of the representation of anchor, positive and negative samples as follows [28]:

$$\| f(x_i^a) - f(x_i^p) \|_2^2 + a < \| f(x_i^a) - f(x_i^n) \|_2^2 \tag{4}$$

where $f(x_i^a)$, $f(x_i^p)$, and $f(x_i^n)$ are the face representations of the anchor, positive, and negative, respectively. All the triplets sampled from the training set should satisfy the constraint. Thus, during training, HaarNet minimizes of the loss function:

$$L_{\text{HaarNet}} = \delta_1 L_{\text{triplet}} + \delta_2 L_{\text{mean}} + \delta_3 L_{\text{std}} \tag{5}$$

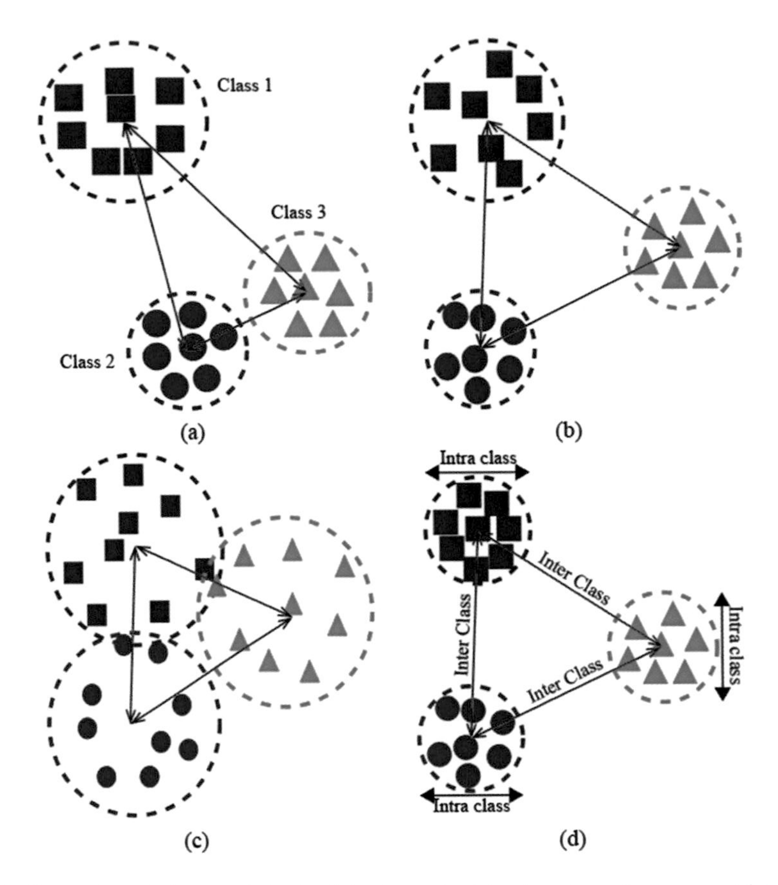

Fig. 9 Illustration of the regularized triple loss principles based on the mean and standard deviation of three classes, assuming 2D representations of the ROIs. (©[2017] IEEE. Reprinted, with permission, from Ref. [25])

where δ_i denotes the weight for each term in the loss function. Furthermore, L_{triplet} can be defined based on (4) as follows:

$$L_{\text{triplet}} = \frac{1}{2N} \sum_{i=1}^{N} \left[\left\| f\left(x_i^a\right) - f\left(x_i^p\right) \right\|_2^2 - \left\| f\left(x_i^a\right) - f\left(x_i^n\right) \right\|_2^2 + \alpha \right]_+ \tag{6}$$

Similar to [8], assuming that the mean distance constraint is $\beta < \left\| \hat{\mu}_c - \hat{\mu}_c^n \right\|_2^2$, L_{mean} is defined as:

$$L_{\text{mean}} = \frac{1}{2P} \sum_{c=1}^{C} \max\left(0,\ \beta - \left\| \hat{\mu}_c - \hat{\mu}_c^n \right\|_2^2\right) \tag{7}$$

In addition, the standard deviation constraint is defined to be $\sigma_c > \gamma$, where σ_c is the standard deviation of the class c. Therefore, L_{std} can be computed as follows:

$$L_{\text{std}} = \frac{1}{M} \sum_{c=1}^{C} \max(0, \gamma - \sigma_c) \tag{8}$$

where N, P, and M are the number of samples that violate the triplet, mean distance, and standard deviation constraints, respectively. Likewise, C is the number of subjects in the current batch, and α, β,and γ are margins for triplet, mean distance, and standard deviation constraints, respectively. The loss function (5) can be optimized using the regular stochastic gradient descent with momentum similar to [8]. The gradient of loss with respect to the facial ROI representation of ith image for subject c (denoted as $f(x_{ci})$) is derived as follows:

$$\frac{\partial L_{\text{std}}}{\partial f(x_{ci})} = -\frac{1}{M} \sum_{c=1}^{C} \omega_c \frac{\partial \sigma_c}{\partial f(x_{ci})} \tag{9}$$

where ω_c equals to 1 if the standard deviation constraint is violated and equals to 0 otherwise. Moreover, the derivative of L_{std} can be computed by applying the chain rule as follows:

$$\frac{\partial \sigma_c}{\partial f(x_{ci})} = \frac{\partial \sqrt{\frac{1}{N_c} \sum_{j=1}^{N_c} \left\| f(x_{cj}) - \mu_c \right\|_2^2}}{\partial f(x_{ci})} = \frac{\left[\sum_{j=1}^{N_c} \frac{1}{N_c} \left\| \mu_c - f(x_{cj}) \right\|_2\right] - \left\| \mu_c - f(x_{ci}) \right\|_2}{2\sqrt{\frac{1}{N_c} \sum_{j=1}^{N_c} \left\| f(x_{cj}) - \mu_c \right\|_2^2}} \tag{10}$$

As shown in Fig. 9d, the discriminating power of the face representations can be improved by setting margins such that $\gamma < \beta$. This ensures a high inter-class and a low intra-class variations to increase the overall classification accuracy.

3.2 Deep CNNs Using Autoencoder

An efficient canonical face representation CNN (CFR-CNN) has been proposed in [26] for accurate still-to-video FR from a SSPP, where still and video ROIs are captured under various conditions. The CFR-CNN is based on a supervised autoencoder that can represent the divergence between the source (still ROI) and target (video ROI) domains encountered in still-to-video FR scenario. The autoencoder network is trained using a weighted pixel-wise loss function that is specialized for SSPP problems and allows to reconstruct canonical ROIs (frontal and less blurred faces) for matching that correspond to the conditions of reference still ROIs. In addition, it can generate discriminative face embeddings that are similar for the same individuals and robust to variations typically observed in unconstrained real-world video scenes. A fully connected classification network is also trained to perform face matching using the face embeddings extracted from the deep autoencoder and accurately determine whether the pairs of still and video ROIs correspond to the same individual.

Autoencoder CNNs are typically utilized to normalize variations in face capture conditions from probe video ROIs to those in still reference ROIs. The architecture of the autoencoder is shown in Fig. 10, where the input image is a probe video ROI captured using a surveillance camera, while the output is a reconstructed image. This network consists of (1) three convolutional layers each followed by a max-pooling layer to extract robust convolutional maps and then (2) a two-layer fully connected network that generates a 256-dimensional face embedding. The decoder reverses these operations by applying a fully connected layer to generate the original vector and three deconvolutional layers, each one followed by un-pooling layers designed for generating the final reconstruction of the face.

A development set (assumed to be collected from unknown individuals captured from the operational domain) is employed for training of the deep autoencoder network. A batch of video ROIs are fed into the network, where still ROIs of the corresponding persons are used for facial reconstructions. Using higher-quality still images that are captured during enrollment under controlled conditions as target faces, the autoencoder network simultaneously learns invariant face embeddings to normalize the input video ROIs. The parameters of this autoencoder network are optimized by employing a weighted mean squared error (MSE) criterion, where a T-shaped region (illustrated in Fig. 11) is considered to assign a higher significance to discriminative facial components like the eyes, nose, and mouth. This loss function of is formulated as:

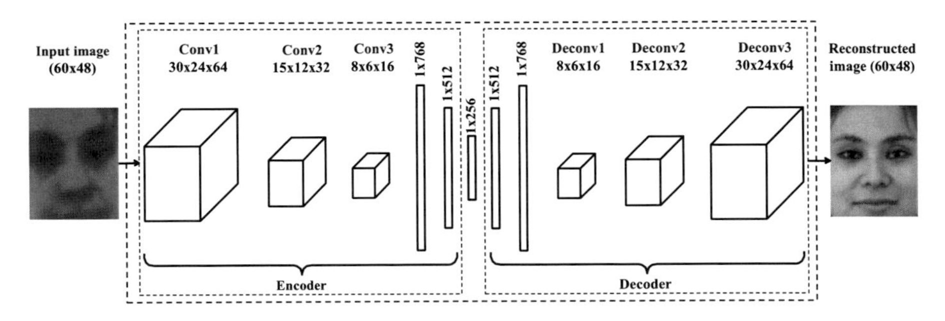

Fig. 10 Block diagram of the autoencoder network in the CFR-CNN. (©[2017] IEEE. Reprinted, with permission, from Ref. [26])

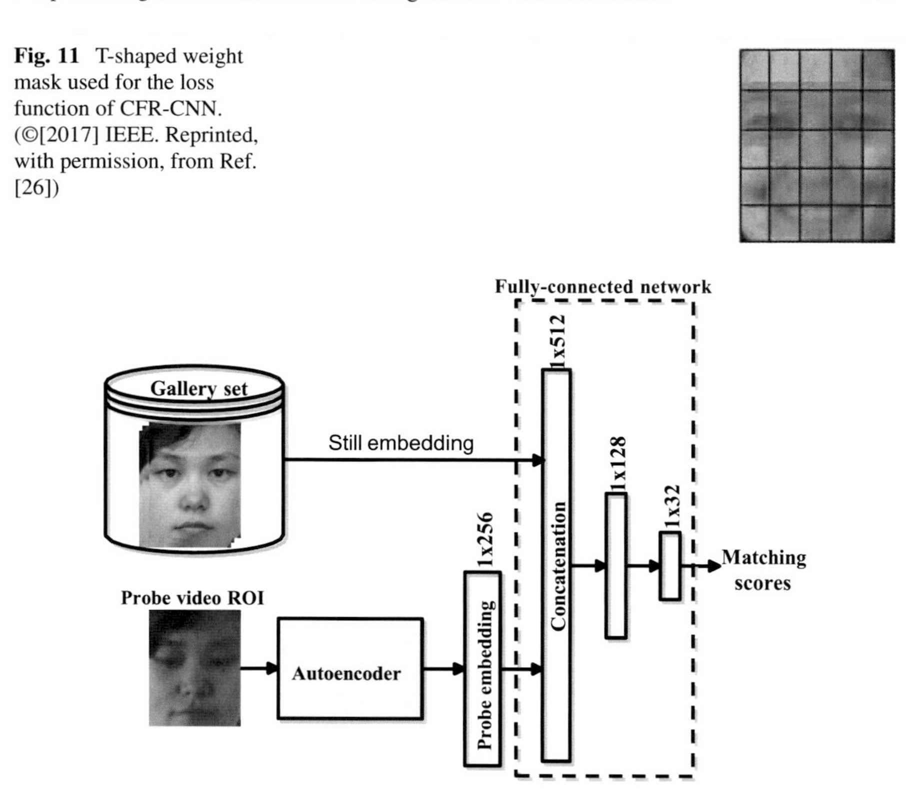

Fig. 11 T-shaped weight mask used for the loss function of CFR-CNN. (©[2017] IEEE. Reprinted, with permission, from Ref. [26])

Fig. 12 Block diagram of the classification network in the CFR-CNN. (©[2017] IEEE. Reprinted, with permission, from Ref. [26])

$$
\begin{aligned}
L_{CFR-CNN} &= \sum_{i \in rows} \sum_{j \in cols} \tau_{i,j} \left\| X^2 - \hat{X}^2 \right\| \\
\tau_{i,j} &= \begin{cases} \alpha \text{ if (i,j) belongs to T} \\ \beta \text{ if (i,j) otherwise} \end{cases}
\end{aligned}
\tag{11}
$$

where $rows \times cols$ is the size of ROIs, X is the target still ROI, and $\hat{X}$ is the reconstructed ROI. The weight α is considered for the T region, while the weight β is considered for pixels outside the T region.

A fully connected network is then integrated with the deep convolutional autoencoder, and the output of the intermediate layer is then considered as a face representation that is invariant to the different nuisance factors commonly encountered in unconstrained surveillance environments. Finally, face matching is performed using a fully connected classification network as shown in Fig. 12. This network is implemented to match the face representations of still and video ROIs.

The fully connected classification network is trained using a regular pairwise matching scheme, where the face embeddings of the reference still and probe video

ROIs are fed into the classification network. The network can thereby learn to classify each pair of still and video ROIs as either matching or nonmatching.

4 Performance Evaluation

The performance of the aforementioned video-based FR systems is evaluated using Cox Face DB [15]. This dataset was specifically collected for video surveillance applications, where it is composed of high-quality still faces captured with still cameras under controlled conditions and low-quality video faces captured with different off-the-shelf camcorders under uncontrolled conditions. Videos are recorded per subject when they are walking through a designed-S curve containing changes in pose, illumination, scale, and blur. An example of still and videos of one subject is shown in Fig. 13.

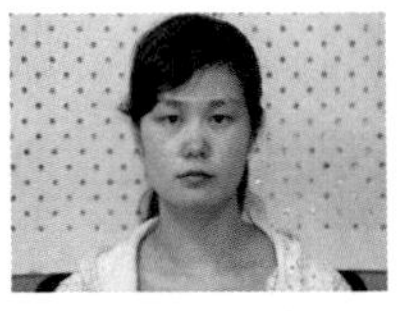

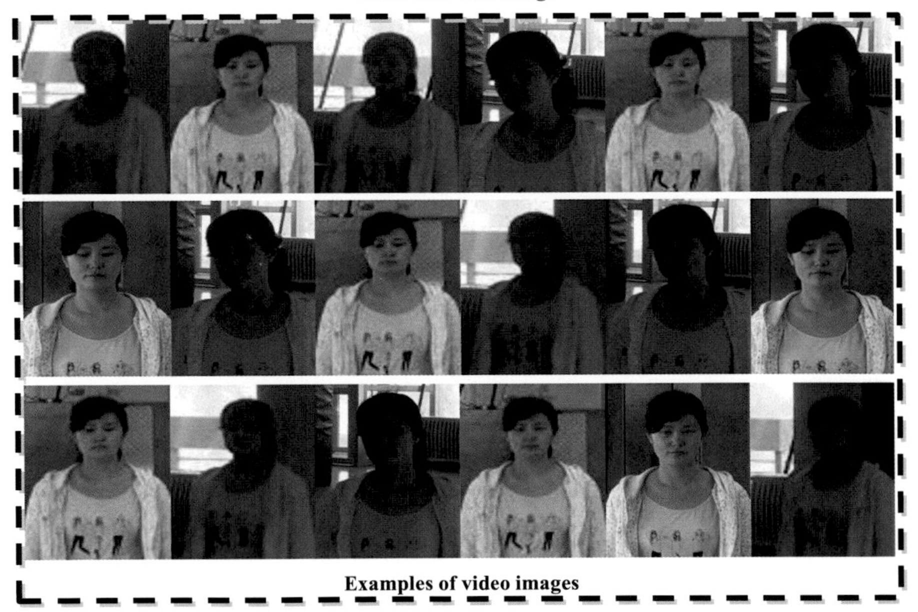

Fig. 13 An example of high-quality reference still image and random low-quality video images of the corresponding individual captured by the still camera and three camcorders in the COX Face DB

Table 1 Rank-1 recognition and computational complexity of video-based FR systems over videos of Cox Face DB

FR system	Rank-1 recognition	Computational complexity		
		# operations	# parameters	# layers
CCM-CNN [24]	89.53 ± 0.9	33.3M	2.4M	30
TBE-CNN [8]	90.61 ± 0.6	12.8B	46.4M	144
HaarNet [25]	91.40 ± 1.0	3.5B	13.1M	56
CFR-CNN [26]	87.29 ± 0.9	3.75M	1.2M	7

The systems are evaluated according to experimental protocol suggested in [15], where each probe video ROI is compared against the reference still ROIs and rank-1 recognition is reported as the FR accuracy. Meanwhile, since video-based FR systems are often required to perform real-time processing in surveillance applications, the computational complexity of such systems should be also taken into consideration. In this regard, the complexity can be determined in terms of the number of operations (to match a video probe ROI to a reference still ROI) and the number of network parameters and layers [5].

In order to confirm the viability of the CNN-based video FR systems for real-time surveillance applications, Table 1 presents the accuracy and compares their computational complexity.

It can be seen in Table 1 that the TBE-CNN and HaarNet provide the highest level of accuracy, while they are very complex. Although the CCM-CNN and CFR-CNN cannot outperform these deep architectures, they can achieve satisfactory results with significantly lower computational complexity. Moreover, the number of network parameters and layers is key factors in designing deep CNN that can greatly affect the convergence and training time. Considering these criteria, the proposed CCM-CNN and CFR-CNN have the lowest design complexity and subsequently the shortest convergence time.

5 Conclusion and Future Directions

In this chapter, the most recent deep learning architectures proposed for robust face recognition in video surveillance were thoroughly investigated. To overcome the existing challenges in real-world surveillance unconstrained environments, the single training reference sample and domain adaptation problems have been taken into account during the system design. On the other hands, computational complexity is also a key issue to provide an efficient solution for real-time video-based FR systems. In particular, this chapter reviewed deep learning architectures proposed based on triplet-loss function and autoencoder CNNs.

Triplet-based loss optimization method allows to learn complex and nonlinear facial representations that provide robustness across inter- and intra-class variations. CCM-CNN proposes a cost-effective solution that is specialized for still-to-video

FR from a single reference still by simulating weighted CCM. TBE-CNN and HaarNet can extract robust representations of the holistic face image and facial components through an ensemble of CNNs containing one trunk and several branch networks. In addition, to compensate the limited robustness of facial model in the case of single reference still, they were fine-tuned using synthetically generated faces from still ROIs of nontarget individuals. In contrast, CFR-CNN employed a supervised autoencoder CNN to generate canonical face representations from low-quality video ROIs. It can therefore reconstruct frontal faces that correspond to capture conditions of reference still ROIs and generate discriminant face representations. Experimental results obtained with the COX Face DB indicated that TBE-CNN and HaarNet can achieve higher level of accuracy with heavy computational complexity, while CCM-CNN and CFR-CNN can provide convincing performance with significantly lower computational costs.

Since the use of deep learning is increasingly growing, one of the future directions is to integrate conventional methods with deep learning methods in order to incorporate statistical and geometrical properties of faces into the deep features. In addition, future research can focus on utilizing temporal information, where facial ROIs can be tracked over frames to accumulate the predictions over time. Thus, the combination of face detection, tracking, and classification in a unified deep learning-based network will lead to a robust spatiotemporal recognition suitable for real-world video surveillance applications. Thus, 3D CNNs and recurrent neural networks such as long short-term memory can be exploited to consider convolutions through the time, due to capturing temporal information among successive video frames.

Acknowledgements This work was supported by the Fonds de Recherche du Québec – Nature et Technologies and MITACS.

References

1. Barr, J.R., Bowyer, K.W., Flynn, P.J., Biswas, S.: Face recognition from video: A review. International Journal of Pattern Recognition and Artificial Intelligence **26**(05) (2012)
2. Bashbaghi, S., Granger, E., Sabourin, R., Bilodeau, G.A.: Watch-list screening using ensembles based on multiple face representations. In: ICPR, pp. 4489–4494 (2014)
3. Bashbaghi, S., Granger, E., Sabourin, R., Bilodeau, G.A.: Dynamic ensembles of exemplar-svms for still-to-video face recognition. Pattern Recognition **69**, 61–81 (2017)
4. Bashbaghi, S., Granger, E., Sabourin, R., Bilodeau, G.A.: Robust watch-list screening using dynamic ensembles of svms based on multiple face representations. Machine Vision and Applications **28**(1), 219–241 (2017)
5. Canziani, A., Paszke, A., Culurciello, E.: An analysis of deep neural network models for practical applications. arXiv preprint arXiv:1605.07678 (2016)
6. Chellappa, R., Chen, J., Ranjan, R., Sankaranarayanan, S., Kumar, A., Patel, V.M., Castillo, C.D.: Towards the design of an end-to-end automated system for image and video-based recognition. CoRR **abs/1601.07883** (2016)
7. Dewan, M.A.A., Granger, E., Marcialis, G.L., Sabourin, R., Roli, F.: Adaptive appearance model tracking for still-to-video face recognition. Pattern Recognition **49**, 129–151 (2016)

8. Ding, C., Tao, D.: Trunk-branch ensemble convolutional neural networks for video-based face recognition. IEEE Trans on PAMI **PP**(99), 1–14 (2017). https://doi.org/10.1109/TPAMI.2017.2700390
9. Gao, S., Zhang, Y., Jia, K., Lu, J., Zhang, Y.: Single sample face recognition via learning deep supervised autoencoders. IEEE Transactions on Information Forensics and Security **10**(10), 2108–2118 (2015)
10. Ghodrati, A., Jia, X., Pedersoli, M., Tuytelaars, T.: Towards automatic image editing: Learning to see another you. In: BMVC (2016)
11. Gomerra, M., Granger, E., Radtke, P.V., Sabourin, R., Gorodnichy, D.O.: Partially-supervised learning from facial trajectories for face recognition in video surveillance. Information Fusion **24**(0), 31–53 (2015)
12. Heo, Y.S., Lee, K.M., Lee, S.U.: Robust stereo matching using adaptive normalized cross-correlation. IEEE Trans on PAMI **33**(4), 807–822 (2011)
13. Huang, G.B., Ramesh, M., Berg, T., Learned-Miller, E.: Labeled faces in the wild: A database for studying face recognition in unconstrained environments. Tech. Rep. 07-49 (2007)
14. Huang, G.B., Lee, H., Learned-Miller, E.: Learning hierarchical representations for face verification with convolutional deep belief networks. In: CVPR (2012)
15. Huang, Z., Shan, S., Wang, R., Zhang, H., Lao, S., Kuerban, A., Chen, X.: A benchmark and comparative study of video-based face recognition on cox face database. IP, IEEE Trans on **24**(12), 5967–5981 (2015)
16. Kamgar-Parsi, B., Lawson, W., Kamgar-Parsi, B.: Toward development of a face recognition system for watchlist surveillance. PAMI, IEEE Trans on **33**(10), 1925–1937 (2011)
17. Kan, M., Shan, S., Su, Y., Xu, D., Chen, X.: Adaptive discriminant learning for face recognition. Pattern Recognition **46**(9), 2497–2509 (2013)
18. Kan, M., Shan, S., Chang, H., Chen, X.: Stacked progressive auto-encoders (spae) for face recognition across poses. In: CVPR (2014)
19. Le, Q.V.: Building high-level features using large scale unsupervised learning. In: ICASSP (2013)
20. Ma, A., Li, J., Yuen, P., Li, P.: Cross-domain person re-identification using domain adaptation ranking svms. IP, IEEE Trans on **24**(5), 1599–1613 (2015)
21. Matta, F., Dugelay, J.L.: Person recognition using facial video information: A state of the art. Journal of Visual Languages and Computing **20**(3), 180–187 (2009)
22. Mokhayeri, F., Granger, E., Bilodeau, G.A.: Synthetic face generation under various operational conditions in video surveillance. In: ICIP (2015)
23. Pagano, C., Granger, E., Sabourin, R., Marcialis, G., Roli, F.: Adaptive ensembles for face recognition in changing video surveillance environments. Information Sciences **286**, 75–101 (2014)
24. Parchami, M., Bashbaghi, S., Granger, E.: Cnns with cross-correlation matching for face recognition in video surveillance using a single training sample per person. In: AVSS (2017)
25. Parchami, M., Bashbaghi, S., Granger, E.: Video-based face recognition using ensemble of haar-like deep convolutional neural networks. In: IJCNN (2017)
26. Parchami, M., Bashbaghi, S., Granger, E., Sayed, S.: Using deep autoencoders to learn robust domain-invariant representations for still-to-video face recognition. In: AVSS (2017)
27. Parkhi, O.M., Vedaldi, A., Zisserman, A.: Deep face recognition. In: BMVC (2015)
28. Schroff, F., Kalenichenko, D., Philbin, J.: Facenet: A unified embedding for face recognition and clustering. In: CVPR (2015)
29. Sun, Y., Wang, X., Tang, X.: Hybrid deep learning for face verification. In: ICCV (2013)
30. Sun, Y., Chen, Y., Wang, X., Tang, X.: Deep learning face representation by joint identification-verification. In: NIPS (2014)
31. Sun, Y., Wang, X., Tang, X.: Deep learning face representation from predicting 10,000 classes. In: CVPR (2014)
32. Sun, Y., Wang, X., Tang, X.: Deeply learned face representations are sparse, selective, and robust. In: CVPR (2015)

33. Szegedy, C., Liu, W., Jia, Y., Sermanet, P., Reed, S., Anguelov, D., Erhan, D., Vanhoucke, V., Rabinovich, A.: Going deeper with convolutions. In: CVPR (2015)
34. Taigman, Y., Yang, M., Ranzato, M., Wolf, L.: Deepface: Closing the gap to human-level performance in face verification. In: CVPR (2014)
35. Vincent, P., Larochelle, H., Lajoie, I., Bengio, Y., Manzagol, P.A.: Stacked denoising autoencoders: Learning useful representations in a deep network with a local denoising criterion. JMLR **11**, 3371–3408 (2010)
36. Yang, M., Van Gool, L., Zhang, L.: Sparse variation dictionary learning for face recognition with a single training sample per person. In: ICCV (2013)
37. Yim, J., Jung, H., Yoo, B., Choi, C., Park, D., Kim, J.: Rotating your face using multi-task deep neural network. In: CVPR (2015)
38. Zheng, J., Patel, V.M., Chellappa, R.: Recent developments in video-based face recognition. In: Handbook of Biometrics for Forensic Science, pp. 149–175. Springer (2017)
39. Zhu, Z., Luo, P., Wang, X., Tang, X.: Multi-view perceptron: a deep model for learning face identity and view representations. In: NIPS (2014)
40. Zhu, Z., Luo, P., Wang, X., Tang, X.: Recover canonical-view faces in the wild with deep neural networks. arXiv preprint arXiv:1404.3543 (2014)

Deep Learning for 3D Data Processing

Zhenbao Liu, Zhizhong Han, and Shuhui Bu

Abstract Extracting local features from raw 3D data is a nontrivial and challenging task that requires carefully designed 3D shape descriptors. In conventional methods, these descriptors are handcrafted and require intensive human intervention and prior knowledge. To tackle this issue, we propose a novel deep learning model, namely, Circle Convolutional Restricted Boltzmann Machine (CCRBM), for unsupervised 3D local feature learning. CCRBM is specially designed for 3D shapes which effectively resolves the obstacles in the hierarchical learning process that existing deep learning models cannot resolve, such as irregular topology of vertices, orientation ambiguity on the 3D surface, and rigid or slightly nonrigid transformation invariance. Specially, by introducing the novel circle convolution, CCRBM holds a novel ring-like multilayer structure to learn 3D local features in a manner of structure preservation. Circle convolution convolves across 3D local regions with a novel circular sector convolution window by rotating itself along a xed circle direction. In the process of circle convolution, extra points are sampled on each 3D local region and projected onto the tangent plane of the center of the region. By this way, the projection distances in each sector window are employed to constitute the raw 3D feature called projection distance distribution (PDD). In addition, to eliminate the ambiguity of the initial location of a sector window, Fourier Transform Modulus (FTM) is used to transform the PDD into Fourier domain which is then conveyed to CCRBM. Experiments using the learned local features are conducted on three aspects: global shape retrieval, partial shape retrieval, and shape correspondence. The experimental results show that the learned local features outperform other state-of-the-art 3D shape descriptors.

This work was supported by the Natural Science Foundation of China under Grant 61672430, 61573284 and 61522207, and NWPU Basic Research Fund under Grant 3102016JKBJJGZ08.

Z. Liu · Z. Han · S. Bu (✉)
Northwestern Polytechnical University, Xi'an, China
e-mail: liuzhenbao@nwpu.edu.cn; h312h@mail.nwpu.edu.cn; bushuhui@nwpu.edu.cn

X. Jiang et al. (eds.), *Deep Learning in Object Detection and Recognition*,
https://doi.org/10.1007/978-981-10-5152-4_7

1 Introduction

Many tasks in 3D shape analysis require the extraction of 3D shape features such as shape retrieval [1], matching [2], segmentation [3], human pose estimation [4], tracking [5], and scene understanding [6]. 3D features can be extracted through global or local shape descriptors. Global shape descriptors are a common approach for shape classification or retrieval that can be considered as a mapping from the space of 3D objects to the vector space with finite dimensional. Comparatively, local shape descriptors are more widely used since they can not only provide local features for point-based shape matching or shape segmentation but also construct global features with the additional global shape description by paradigms such as Bag-of-Words [7], spatial-sensitive Bag-of-Features [8]. However, almost all proposed shape descriptors are handcrafted [9–11] that involves labor-intensive design, human ingenuity, and prior knowledge.

Although local shape descriptors can effectively capture the discriminative local or even global geometric characteristics, their manual tune-up procedure is time-consuming and normally does not generalize well, which burdens the extraction and organization of the discriminative information from 3D shapes when constructing classifiers or other predictors [12]. In addition, existing local descriptors are only based on a single local region, such that the learned features are highly sensitive on local changes and hence an invariant pattern of features cannot be easily and stably obtained. From these viewpoints, it is significant and promising to explore unsupervised feature learning methods to learn a common and invariant pattern of 3D local features, especially when more and more 3D objects are publicly available.

Feature learning has been attracting more and more research interests [12], which includes two main categories: supervised feature learning [13] and unsupervised feature learning [14]. Since unsupervised feature learning does not require training labels, it is more advantageous over supervised feature learning. Among the various recent unsupervised feature learning methods, deep learning model is the most popular for its attractive characteristics:

1. High-level and hierarchical representations can be learned via multiple nonlinear transformations of the data.
2. Feature learning process can be performed in unsupervised way.

Many practical applications including computer vision and natural language processing [15–17] benefit from deep learning models. However, the existing deep learning models cannot help 3D shape analysis to enhance performance since they are impossible to directly learn high-level features for 3D shapes. Several obstacles are listed as follows:

1. The topology of a 3D mesh is irregular. Since a mesh is not a regular lattice as a 2D image, it is impossible to feed the whole or part of mesh into deep learning models which contain only regular structure.

2. The orientation is ambiguous on the surface of 3D shapes. Some convolution-based deep learning methods cannot convolve across the 3D shape surface as the same way of a 2D image.
3. A 3D shape can always be transformed rigidly and nonrigidly through various transformations such as translation, rotation, and bending. Therefore, a learning model is required to work invariantly on these rigid and slightly nonrigid transformations.

An intuitive way of overcoming these obstacles is to project 3D shapes into a 2D intermediate representation space by handcrafted feature descriptors [18, 19] or view-based methodology [20]. In the representation space, deep learning models can be easily applied. However, the transformation of 2D representation for 3D shapes definitely incurs a significant amount of information loss. For example, the structural information of a 3D surface, which is exactly the learning target by deep learning models, shall be lost in the 2D representation space. In addition, the view-based methodology can only learn global features but cannot learn local features, and methods based on the handcrafted features cannot get rid of the parameter Ąne-turning procedure before learning by deep learning models.

To alleviate these problems, we propose a novel deep learning model for unsupervised learning of high-level discriminative local features from raw representation of 3D shapes, namely, *Circle convolutional restricted Boltzmann machines* (CCRBM). Based on the learning framework of *convolutional restricted Boltzmann machines* (CRBM) [15], CCRBM holds a novel ring-like multilayer model structure to learn the geometric and structural information of 3D local regions by a number of convolution filters via a new convolution manner called circle convolution. The circle convolution moves a novel type of convolution window along a fixed circle direction on a surface, which not only overcomes the orientation ambiguity on a 3D surface but also sequentially captures the geometric and structural variations of the surface. Specifically, the sampled points from a given local region are projected onto the tangent plane of the center of the region. By this way, the projection distances in each sector window are employed to constitute the raw 3D feature called *projection distance distribution* (PDD) which is only determined by the coordinates of the sampled vertices. Another issue is the initial but usually ambiguous location of the convolution window on the 3D surface. This issue can be eliminated by introducing *Fourier Transform Modulus* (FTM) which transforms the PDD into an invariant representation of the initial location for CCRBM. Moreover, circle convolution can work directly on a 3D local surface, which resolves the issues of feature invariance for rigid and slightly nonrigid transformation.

The paper is organized as follows: Sect. 2 presents the related works of handcrafted 3D local features and preliminaries about deep learning models for feature learning. The designs of circle convolution, PDD, and CCRBM are detailed in Sect. 3. Experimental setup and results with analysis are shown in Sects. 4 and 5, respectively. Finally a conclusion with our contributions is drawn in Sect. 6.

2 Related Works

Existing 3D local features are mainly extracted by knowledge-based descriptors. Recently, neural network-based methods are involved in learning global features. These methods are briefly reviewed in the following subsections.

2.1 Knowledge-Based 3D Local Features

Many knowledge-based local descriptors have been proposed for 3D shapes [21]. Some local descriptors extract the property of a local region via a scalar such as curvature [22] and volume [9]. Another kind of local descriptors captures the relative spatial information such as spin image [10] and shape contexts [11]. In nonrigid shape analysis, some intrinsic local descriptors are proposed based on diffusion geometry method such as heat kernel signature (HKS) [23], scale-invariant heat kernel signature (SIHKS) [24], and Wave Kernel Signature (WKS) [25].

Although effective, these knowledge-based local descriptors are all handcrafted, and heavy human intervention with specific knowledge is required to tune the descriptor parameters for satisfactory results. Moreover, these descriptors can merely extract features from a single region but not a common pattern of features from a set of similar regions, leading to oversensitivity of small changes and degeneration of discriminability. In this paper, the proposed CCRBM aims to alleviate all of these issues.

2.2 Deep Learning for 3D Shapes With Raw Features

Deep learning model is a kind of neural network for unsupervised feature learning. There have been some works on learning features from raw features for 3D shapes using deep learning methods. In the literature [26], a model based on a combination of convolution and recursive neural networks (CNN and RNN) was introduced for learning features of RGB-D images and classifying the corresponding 3D objects. However, their model is different from ours in the way that global features are constructed by merging the information learned separately from 2D RGB images and depth images. In this way, their model cannot learn the local features of a real 3D object. Moreover, the structural information of the 3D object is lost when learning 3D features using single view of 2D images only. In CCRBM, local features on 3D surfaces are directly learned in a structure preserving manner, which tries to preserve as much information as possible when mapping 3D local regions to the feature space.

3D ShapeNets [27] is a specially designed deep learning model for learning global 3D shape features. In 3D ShapeNets, the representations of a geometric 3D shape are regarded as a probability distribution of binary variables on a 3D voxel grid. In addition, each 3D mesh is treated as a 48 * 48 * 48 binary tensor: 1s indicate the voxels on or inside the mesh surface, and 0 s indicate the voxels outside the mesh. The difference between our work and 3D ShapeNets is threefold.

1. The representation of training shapes is different. For an input 3D object represented as mesh, 3D ShapeNets has to voxelize the object beforehand, while CCRBM does not require any data preprocessing and can directly learn features from local regions.
2. CCRBM learns features directly on the surface of 3D shapes. Through this way, the effect of the rigid transformation (e.g., translation and rotation) and slightly nonrigid transformation (e.g., bending) can be minimized. However, considering the essence of voxelization, the input objects of 3D ShapeNets must be rigid shapes (e.g., table and chair) rather than nonrigid ones (e.g., human and animal); moreover, the shapes have to be aligned in the same direction before voxelization, which is however a very difficult task for current alignment methods. Thus, it is hard for 3D ShapeNets to withstand the effect of the rotation and the nonrigid transformations of input shapes.
3. CCRBM learns 3D local features, which can be used to construct global features via Bag-of-Words framework as well, while 3D ShapeNets learns only 3D global features.

2.3 Restricted Boltzmann Machines (RBM) and Deep Belief Network (DBN)

In the literature, RBM and DBN were proposed to ease the optimization difficulty associated with deep learning models by the idea of greedy unsupervised pre-training of each stacked layer [28, 29]. An RBM is a generative stochastic neural network that can learn a probability distribution over its set of inputs. As the name implies, an RBM is a variant of Boltzmann machines [30], with the restriction that each connection in an RBM must connect a visible node to a detection node and there is no connection between nodes in the same layer as shown in Fig. 1a. This restriction leads to efficient training algorithms, in particular the gradient-based contrastive divergence algorithm [28, 31]. A DBN is constructed by stacking several RBMs layer-by-layer, the lower layer detects simple features, and then it is conveyed into higher layers, which in turn detects more complex features.

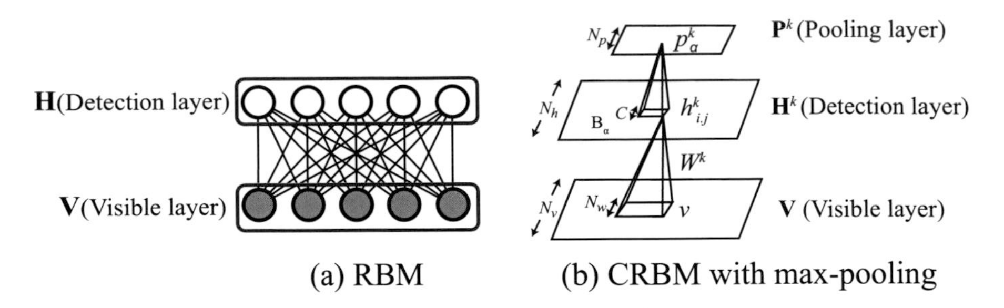

Fig. 1 (**a**) The model of RBM. (**b**) The model of CRBM with max-pooling

2.4 *Convolutional RBM (CRBM) and Convolutional DBN (CDBN)*

To make RBM scaled gracefully for high-dimensional images and obtain local translation-invariant representation, Lee et al. [15] proposed *Convolutional RBM* (CRBM) and *Convolutional DBN* (CDBN). A CRBM is an extension of the traditional RBM to a convolutional setting that aims to learn statistical relationship between a visible layer **V** and a detection layer **H** which contains K "groups" of units as shown in Fig. 1b. A max-pooling layer **P** combined with CRBM shrinks the detection layer using a pooling window B_α of C pixel width. The learned feature detectors W^k ($k \in [1, K]$) are shared among all locations in an image, since features capturing useful information in one region can pick up the same information in the other regions. Similar to DBN, a CDBN can also be constructed by stacking several CRBMs together. To make our paper self-contained, the details of CRBM and CDBN are briefly reviewed. Experienced readers can also jump to Sect. 3 for the proposed CCRBM. CRBM is in fact a building element of CDBN. A CRBM consists of two layers: a visible layer **V** and a detection layer **H** as shown in Fig. 1b. The detection nodes are binary-valued, and the visible nodes are binary-valued or real-valued.

2.5 *Details of CRBM and CDBN*

To make our paper self-contained, the details of CRBM and CDBN are briefly reviewed. Experienced Readers can also jump to Sect. 3 for the proposed CCRBM. CRBM is in fact a building element of CDBN. A CRBM consists of two layers: a visible layer **V** and a detection layer **H** as shown in Fig. 1b. The detection nodes are binary-valued, and the visible nodes are binary-valued or real-valued.

Consider the visible layer **V** consisting of an N_v dimensional array of binary nodes as shown in Fig. 1b. K filter weights are set, and each filter is a N_w dimensional array. For k-th filter W^k, it convolves across an image to obtain the

k-th group $\mathbf{H}^k$ of the detection layer. The $\mathbf{H}^k$ is an N_h dimensional array with nodes h^k_{ij} sharing the same weights W^k and a same bias b^k, where i, j denotes the vertical and horizontal index of the array. All the nodes v_{ij} in the visible layer share the same bias c.

The probabilistic semantics $P(\mathbf{v}, \mathbf{h})$ over visible nodes $\mathbf{v}$ and detection nodes $\mathbf{h}$ is defined in (1), where $Z = \sum_{\mathbf{v}} \sum_{\mathbf{h}} exp(-E(\mathbf{v}, \mathbf{h}))$ is a normalization factor or partition function.

$$P(\mathbf{v}, \mathbf{h}) = \frac{1}{Z} exp(-E(\mathbf{v}, \mathbf{h})) \tag{1}$$

The energy function of CRBM can then be defined in (2).

$$E(\mathbf{v}, \mathbf{h}) = -\sum_{k=1}^{K} h^k \bullet (\widetilde{W}^k * v) - \sum_{k=1}^{K} b^k \sum_{ij} h^k_{ij} - c \sum_{ij} v_{ij} \tag{2}$$

$*$ and $\bullet$ are used to denote 2D convolution and the element-wise product followed by summation, respectively. The tilde above W^k is used to denote flipping W^k horizontally and vertically. Since all nodes in each layer are conditionally independent given the other layer, block Gibbs sampling [32] can be used to perform inference in the network efficiently.

The conditional probability distributions used by Gibbs sampling can be defined as (3) and (4), where the sigmoid function is defined as $sigmoid(x) = \frac{1}{(1+exp(-x))}$.

$$P(h^k_{ij} = 1|\mathbf{v}) = sigmoid((\widetilde{W}^k * v)_{ij} + b_k) \tag{3}$$

$$P(v_{ij} = 1|\mathbf{h}) = sigmoid(\sum_{K} (W^k * h^k)_{ij} + c) \tag{4}$$

In order to learn high-level representations, we can stack CRBMs into a multi-layer architecture analogous to DBNs, and this architecture is based on probabilistic max-pooling. Lee et al. [15] further combined CRBM with probabilistic max-pooling where the maximum over small neighborhoods of detection nodes h^k_{ij} are computed probabilistically. Then, the energy function of the probabilistic max-pooling CRBM can be defined in (5).

$$\begin{aligned} E(\mathbf{v}, \mathbf{h}) = -\sum_{K} \sum_{ij} (h^k_{ij} (\widetilde{W}^k * v)_{ij} + b_k h^k_{ij}) - c \sum_{ij} v_{ij} \\ s.t. \sum_{(ij) \in \mathrm{B}_\alpha} h^k_{ij} \leq 1, \forall k, \alpha \end{aligned} \tag{5}$$

where the B_α defines a small region α on the detection layer as a pooling window (C-pixel width).

In principle, the CRBM parameters can be optimized by performing stochastic gradient ascent on the log-likelihood of training data. However, computing the exact gradient of the log-likelihood is intractable. Thus, contrastive divergence approximation [31] is always used in practice.

Specifically, given the visible layer $\mathbf{V}$, the detection layer $\mathbf{H}^k$ and the pooling layer $\mathbf{P}^k$ can be sampled. With the k-th filter, the corresponding map receives the following bottom-up signal from layer $\mathbf{V}$ as (6).

$$I(h_{ij}^k) \triangleq (\tilde{W}^k * v)_{ij} + b_k \tag{6}$$

Suppose $(i, j) \in \mathrm{B}_\alpha$, the energy increased by tuning on h_{ij}^k is $-I(h_{ij}^k)$, and the conditional probability is defined as the following.

$$P(h_{ij}^k = 1|\mathbf{v}) = \frac{exp(I(h_{ij}^k))}{1 + \sum_{(i'j') \in \mathrm{B}_\alpha} exp(I(h_{i'j'}^k))} \tag{7}$$

$$P(p_\alpha^k = 0|\mathbf{v}) = \frac{1}{1 + \sum_{(i'j') \in \mathrm{B}_\alpha} exp(I(h_{i'j'}^k))} \tag{8}$$

As we can see, (7) and (8) used by Gibbs sampling form the basis of CRBM inference and learning algorithms.

CDBN is constructed by stacking several max-pooling CRBMs. Training CDBN is achieved in a greedy and layer-wised manner: once a given layer is trained, its weights are fixed, and its activations are used as input to the next layer. Our CCRBM is a variation of CRBM. Consequently, a CCDBN can be constructed in the same way of constructing a CDBN by stacking several probabilistic max-pooling CCRBMs.

3 Circle Convolutional RBM (CCRBM)

Before introducing the structure of CCRBM, we first explain the circle convolution and the raw 3D feature called PDD. The resolution of ambiguity of initial location of a sector window used by circle convolution is also discussed.

3.1 Circle Convolution

Introduction Circle convolution is specially designed for performing convolution on a 3D surface which defines both a novel type of convolution window and a novel window moving manner. By performing circle convolution, the structure of a local region can be captured by scanning the local region window-by-window sequentially.

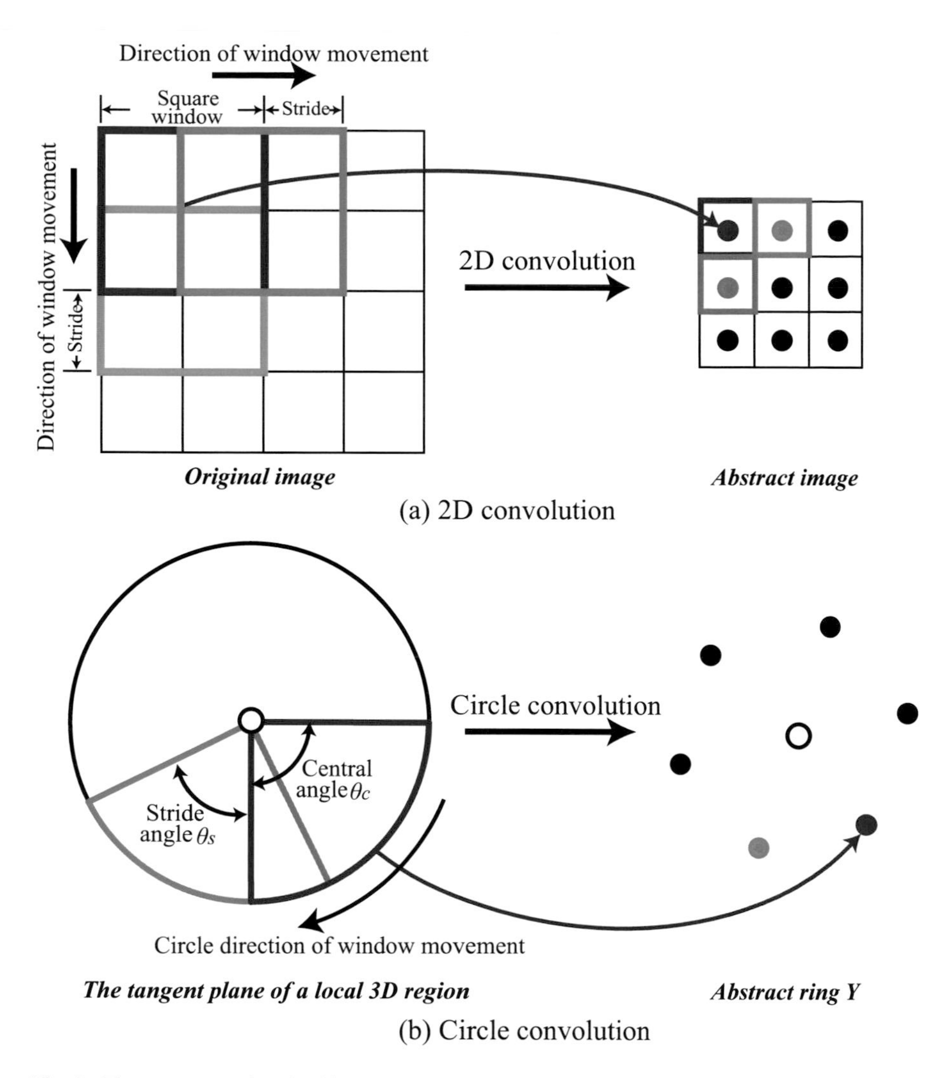

Fig. 2 The parameters involved in 2D convolution and circle convolution

Due to the regular lattice as 2D images, the 2D convolution convolves across an image by moving a rectangle or a square convolution window step-by-step, as shown in Fig. 2a. In this example, the size of a convolution window in the original image is 2×2 pixel. The moving stride along the horizontal and vertical direction is both one pixel. The initial location of the convolution window is marked as the red square. After a filter convolves with the color in that window, the obtained value is conveyed to the corresponding red node in an abstract image on the right as the red arrow shows. Then, the convolution window moves along the horizontal direction by one stride (one pixel in this example) to the location marked as the blue

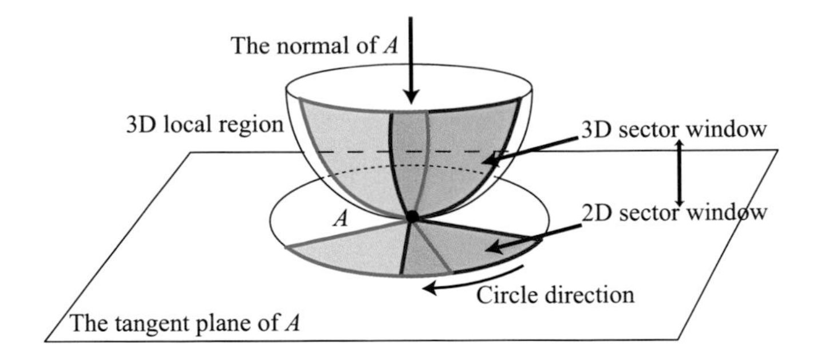

Fig. 3 An illustration of circle convolution on a 3D local region

square. Accordingly, the convolution result is stored in the blue node in the abstract image. The same process can be repeated by moving the convolution window along the vertical direction, just shown as the green square and the corresponding green node in the abstract image. To alleviate the difficulty brought by irregular connection between points and orientation ambiguity on the surface, we propose circle convolution to convolve across a 3D local surface.

Circle convolution adopts a circular sector window to convolve across a local region by rotating the window about the normal of the center point on the local region. As shown in Fig. 3, a red sector window rotates about the normal of the center point A to the position marked in blue on the 3D local region. Note that there could be some overlapping area covered by the two successive sector window locations. The rationale to design the circle convolution is based on the modeling of a 3D local region, which can be modeled as an approximate circular patch centered at a point A with a given radius R. Then, a circular sector window can cover a part of the patch. When rotating about the normal of the center point A sequentially, the sector window can scan the whole surface of the 3D local region.

Circle convolution window approximation However, it is nontrivial to directly and accurately draw a circular sector window on a 3D local region, since the surface is not necessarily a plane and the connection topology or spatial relations between vertices is irregular. Thus, we resort to the tangent plane of the center point of the local region to approximate a 3D circular sector window. In Fig. 3, the sampled points from the 3D local region are firstly projected onto the tangent plane. Then, a 2D circular sector window can be easily established on the tangent plane, such as the one enclosed by two radii and an arc in red in Fig. 3. The sampled points whose projections are covered by the red 2D sector window correspond to a region on the 3D surface. This region can be regarded as the one covered by an approximate 3D sector window, also shown in red and with red boundary on the surface in Fig. 3. When the red 2D sector window rotates to the blue sector window about the normal on the tangent plane, the corresponding red 3D sector window moves to the location marked in blue via rotating about the same axis and along the same direction accordingly.

The movement direction of a circle convolution window The circle direction of a sector movement is defined under the famous right-hand corkscrew-rule. Simply speaking, a 2D sector window is moving about a directed axis along the direction of wrapping one's right hand when the thumb is in the same direction of the axis. In Fig. 3, the red circular sector window rotates anticlockwise when the view is in the negative direction of the normal. Therefore, circle convolution defines a fixed moving direction of the sector window which overcomes the issue of the orientation ambiguity on the surface.

The parameters of circle convolution There are two parameters of circle convolution that specify the size of a sector window and the stride of the rotation, respectively. Similarly, it is hard to parameterize a 3D sector window and its movement on a 3D local region exactly. Alternatively, the 2D sector window projected on the tangent plane is used to parameterize the corresponding 3D sector window, since the size and the movement of the 2D sector window are synchronized with the 3D sector window. These parameters for a 2D sector window involved in circle convolution are illustrated in Fig. 2b. One parameter is the central angle of a 2D circular sector window θ_c, which controls the coverage of the 2D sector window and its corresponding 3D sector window. Another parameter is the stride angle θ_s, which represents the rotation angle of the 2D sector and its corresponding 3D sector window along the circle direction.

Circle convolution After defining the 3D sector window and its movement, we formulate the circle convolution as (9).

$$Y(\theta_s) = \int_{\theta=0}^{2\pi} h(\theta_s - \theta) X(\theta) d\theta \tag{9}$$

where X denotes raw features on the 3D patch and h is a filter to convolve across the 3D patch. By circle convolution, an abstract ring Y composed of nodes whose locations are determined by θ_s is obtained as shown in the right of Fig. 2b.

When h convolves with X extracted from a specified 3D sector window (e.g., the one corresponding to the red 2D sector window in Fig. 3), Y will be stored in the red node (the red arrow demonstrates this process). Then, h moves to the location marked in blue. The aforementioned process is repeated again. Each node in Y corresponds to a location of the rotating 2D sector window. By this way, Y encodes the geometric information of the 3D patch in a structure preserving manner.

Theoretically, the essence of circle convolution lies in abstracting information from raw features by convolution as (9). Through the convolution process, the learned filters h will highlight different particular feature patterns at explicit locations determined by θ_s. In addition, it contains novel convolution window, movement of the window, and parameterization which can work with the irregular vertex topology and ambiguous orientation on the 3D surface.

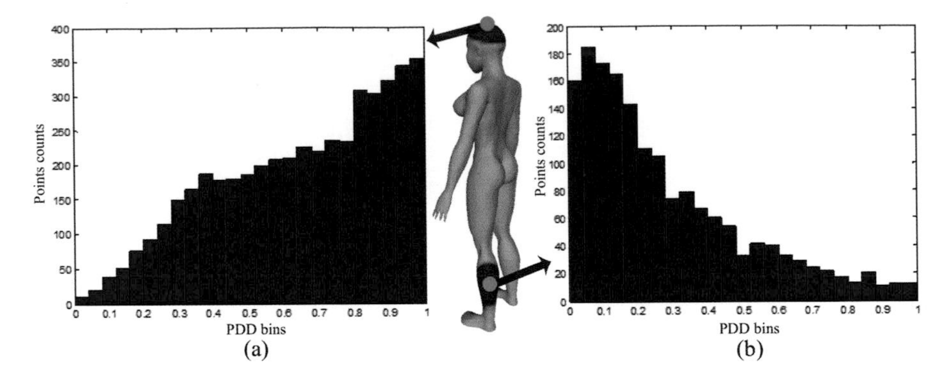

Fig. 4 The projection distance distribution (PDD) of a region centered at (**a**) a point on the human head, (**b**) a point on the human leg

3.2 Projection Distance Distribution (PDD)

The geometric information of a 3D region covered by a sector convolution window can be captured as a PDD via projecting all points in the window to the tangent plane of the region's center. A PDD is a 1D histogram that counts the number of points in each projection distance bin, which can encode different geometric properties of different regions. As shown in Fig. 4, the sampled points in the red region centered at the green point on the human head produce a PDD, which is completely different from the PDD of the region centered at the blue points on the leg. However, the number of projection distance bins affects the discriminability of PDD. Too many bins would present too much details of the distribution which hides the underlying distribution trend. On the contrary, if there were not enough bins, the PDD would be too coarse to capture discriminative information. Through experiments, a number of 25 bins are found suitable to construct a PDD vector in each sector window.

PDD is the raw 3D feature which does not require any handcrafted features but just the coordinates of points on the local region. CCRBM learns features from PDD by circle convolution, which accords with the essence and intention of unsupervised feature learning by deep learning method. In current experiments, as a trade-off between discriminability and computational cost of PDD, we randomly sample 50 points on each triangle face by following the method [33].

3.3 Example of Circle Convolution and PDD Computation

An example of circle convolution and PDD computation is shown in Fig. 5. In order to alleviate the computation burden, CCRBM learns some of local regions centered at sampled points on each shape. For the ant model shown in Fig. 5a, 500 points (in blue) are sampled using farthest geodesic sampling (FGS) method [34]. FGS first

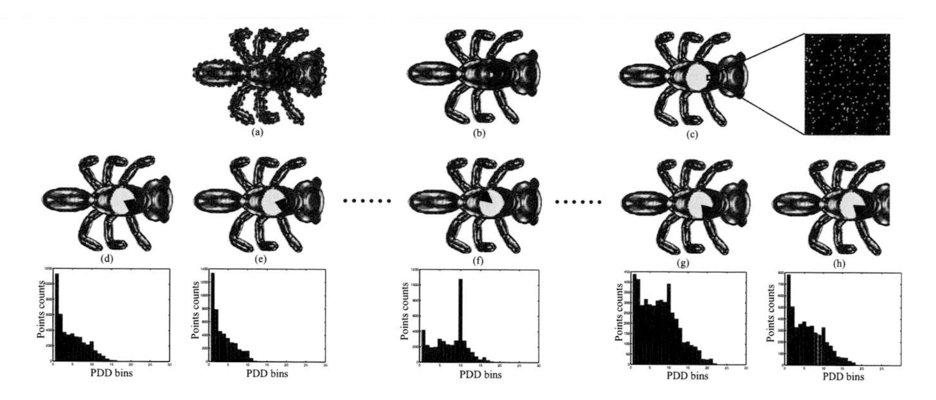

Fig. 5 An example of circle convolution and PDD computations. (**a**) Sampled vertices to learn. (**b**) A local region centered at a sampled vertex. (**c**) More sampled points on triangle faces. (**d**) Sector window 1 and its PDD. (**e**) Sector window 2 and its PDD. (**f**) Sector window 20 and its PDD. (**g**) Sector window 35 and its PDD. (**h**) Sector window 36 and its

samples an arbitrary vertex. The second vertex to be sampled is the one with the farthest geodesic distance to the first sampled one. Note that geodesic distance is the distance of the shortest path connecting two vertices on the surface. Subsequently, the third sampled vertex is the farthest vertex to the first two sampled vertices. The sampling process is repeated until the number of sampled vertices reaches 500.

For every blue local region centered at the sampled points (e.g., the yellow point in Fig. 5b), a more discriminative PDD is computed, as described in Sect. 3.2. In Fig. 5c and its magnified views, additional 50 more points are sampled on each triangle face covered by the blue local region, which are marked as yellow points. Then, a 3D sector window is determined as described in Sect. 3.1. The first sector window with $\theta_c = 60°$ (in BLACK) and the corresponding PDD are shown in Fig. 5d. According to the right-hand corkscrew-rule, the sector window rotates about the normal at the center along the anticlockwise direction (view in the negative direction of the normal at the center) to the next location as shown in Fig. 5e. The procedure repeats and Fig. 5f–h shows the other subsequent steps of the sector window and their corresponding PDDs.

3.4 Elimination of the Initial Location Ambiguity

The ambiguity of the initial location of a sector window is another significant issue. Specifically, for a given 3D local region, the computed PDDs of all sector windows can be seen as a function of the initial location θ, the sector index j_f and the stride step θ_s, denoted as $f_{PDD}(\theta + j_f \times \theta_s)$, where $j_f \in \{0, 1, \ldots, 360°/\theta_s - 1\}$. Remark: actually, in this polar coordinate alike frame, the scalar variable θ cannot parameterize the initial location of a sector window since there is no polar axis which is in a fixed direction. However, θ could help to represent the relationship between two initial locations. Obviously, different initial location θ over the same 3D local region produces a different f_{PDD}. Since there is no consistent local coordinate system on a surface, it is impossible to obtain an identical f_{PDD} on a 3D local region via determining an invariant initial location of a sector window. On the other hand, for two similar local regions, the different initial location-dependent PDDs of each local region will also make CCRBM impossible to extract similar information. In this study, Fourier Transform Modulus (FTM) is proposed to eliminate this ambiguity of the initial location.

FTM has been widely used in image registration [35] and was introduced for the construction of scale- and rotation-invariant descriptors in [36]. It is also used for constructing scale-invariant heat kernel signatures (SIHKS) [24] and intrinsic shape context [37] for 3D shapes. In our work, f_{PDD} is transformed into spectral domain with respect to the angular location $\theta + j_f \times \theta_s$ by FTM. The representation of f_{PDD} in spectral domain is denoted as $F_{j_f}\{f_{PDD}(\theta + j_f \times \theta_s)\}(\omega)$, where ω represents the frequency. The elimination of initial location ambiguity is explained as follows. Setting $\theta = \beta$, the representation of f_{PDD} in spectral domain is obtained by FTM as shown in (10).

$$F_{j_f}\{f_{PDD}(\beta + j_f \times \theta_s)\}(\omega) = \sum_{j_f} f_{PDD}(\beta + j_f \times \theta_s) exp(-i\omega(\beta + j_f \times \theta_s)) \tag{10}$$

If the PDD is obtained from another initial location which is represented by adding an offset δ to β, such that $\theta = \beta + \delta$, the transformation of FTM is shown in (11).

$$F_{j_f}\{f_{PDD}(\beta + \delta + j_f \times \theta_s)\}(\omega) = F_{j_f}\{f_{PDD}(\beta + j_f \times \theta_s)\}(\omega) exp(-i\delta\omega) \tag{11}$$

Taking the modulus, we have (12). Thus, the FTM of PDD, $|F_{j_f}\{f_{PDD}(\theta + j_f \times \theta_s)\}(\omega)|$, is an invariant representation of the initial location for a 3D local region which will be conveyed into the virtual layer as the input of CCRBM, as discussed in Sect. 3.5.

$$|F_{j_f}\{f_{PDD}(\beta + j_f \times \theta_s)\}(\omega)| = |F_{j_f}\{f_{PDD}(\beta + \delta + j_f \times \theta_s)\}(\omega)| \tag{12}$$

Figure 6 shows the elimination of the ambiguity of the initial location. In Fig. 6a, b, a black sector window with different initial locations convolves across a same local region covered by sampled yellow points, respectively. f_{PDD} and $f_{PDD}1$ obtained with two different initial locations are different, as shown in Fig. 6c, d. After performing FTM with respect to the location angle, $|F_{j_f}\{f_{PDD}\}|$ and

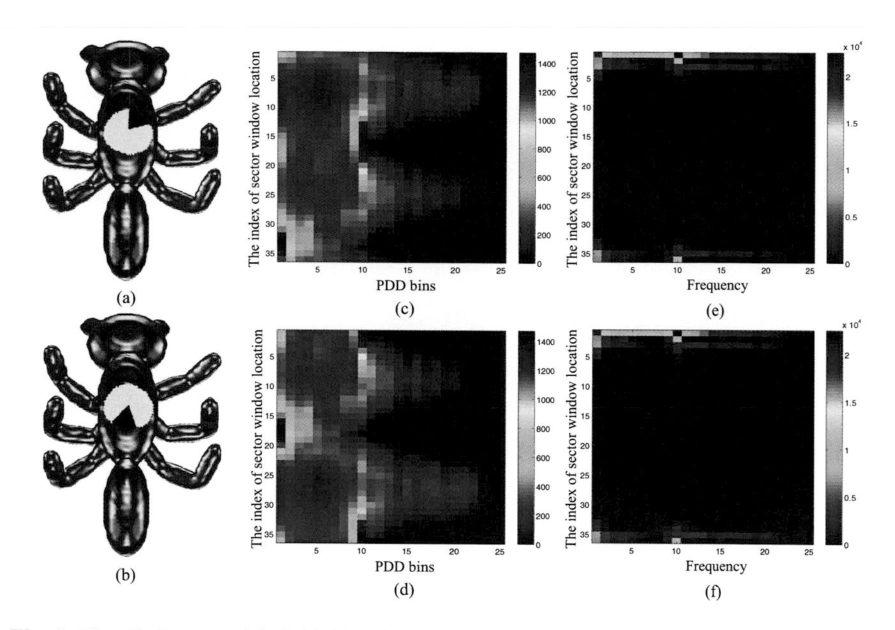

Fig. 6 The elimination of the initial location of the sector window by FTM. (**a**) One initial location of a sector window. (**b**) Another initial location of a sector window. (**c**) f_{PDD} (**d**) $f_{PDD}1$ (**e**) $|F_{j_f}\{f_{PDD}\}(\omega)|$ (**f**) $|F_{j_f}\{f_{PDD}1\}(\omega)|$

$|F_{j_f}\{f_{PDD}1\}|$ are obtained, and they are the same as shown in Fig. 6e, f. From Fig. 6, we can see the ambiguity of the initial location of the sector window makes the f_{PDD} uncertain; however, $|F_{j_f}\{f_{PDD}\}|$ eliminates the ambiguity and presents an initial location-invariant representation to the CCRBM.

3.5 The Structure of CCRBM

In Fig. 7, CCRBM with max-pooling consists of three layers, namely, virtual layer **V**, detection layer **H**, and pooling layer **P**. The significant difference between CCRBM and CRBM is the topology of nodes in each layer, where the nodes in CCRBM layers are lined up in a ring rather than 2D lattice as CRBM. In the visible layer **V1**, the f_{PDD} of a local region with N stride steps are computed, where $N = 360°/\theta_s$. Note that the PDD extracted from each sector window is a m-dimensional vector, where $m = 25$ (i.e., number of bins) in all our experiments. Then, the initial location-invariant representations of f_{PDD} obtained by FTM are conveyed into the $N \times m$ nodes in virtual layer **V**, where **V** acts as the input of CCRBM.

The detection and pooling layers both have K groups of nodes. For each $k \in [1, K]$, the responses of a ring detection layer $\mathbf{H}^k$ with N nodes are generated by circle convolution via convolving with a $1 \times m$ filter W^k across the virtual layer. The

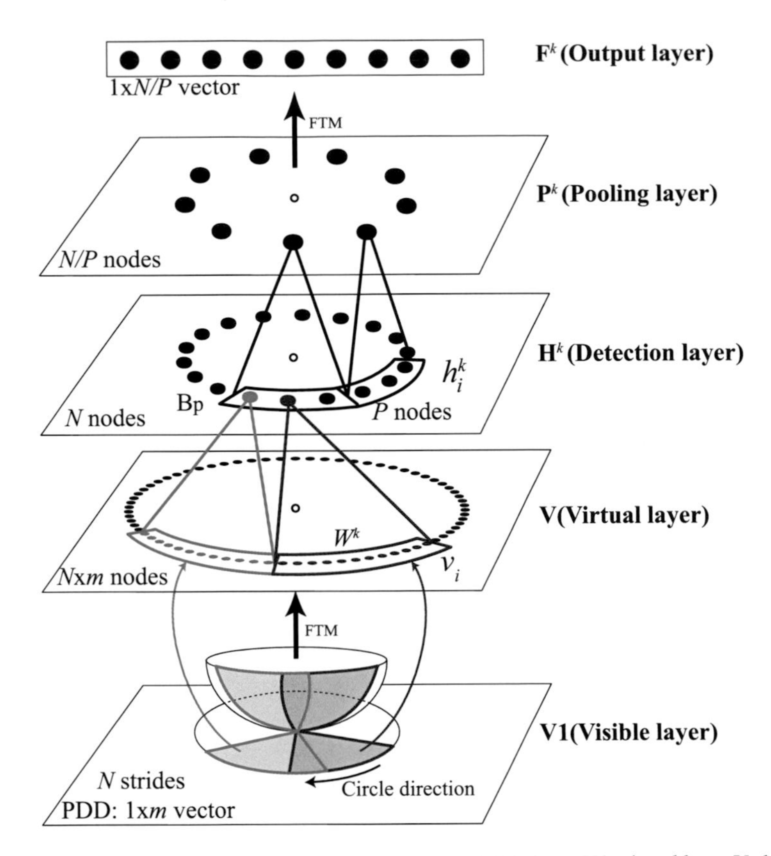

Fig. 7 The structure of max-pooling CCRBM including visible layer V1, virtual layer V, detection layer **H**, pooling layer **P** and output layer **F**

filter convolves with each FTM of PDD in the virtual layer sequentially. This process is shown briefly in the virtual layer in Fig. 7, in which a filter firstly convolves with nodes in the red window. The convolution result is conveyed to a node in the detection layer which is also marked in red. Secondly, the red window rotates with striding m nodes to the location marked in blue. The filter will convolve with nodes in the blue window as well, and the convolution result is conveyed to the blue node in the detection layer accordingly.

The pooling layer $\mathbf{P}^k$ shrinks the representation of the detection layer $\mathbf{H}^k$ by selecting the maximum activations among P adjacent nodes (P-node pooling window) covered by a pooling window B_p, and each pooling layer $\mathbf{P}^k$ has N/P nodes. The shrinking of the representation with max-pooling provides two benefits. Firstly, the representations in higher layer become invariant subject to small fluctuation of the response in the lower layer. Secondly, computational complexity is also reduced in the subsequent procedure.

The output layer is obtained by transforming the pooling layer via FTM. When the nodes in a ring in the pooling layer are stretched to be a N/P-dimensional vector of feature $\mathbf{F}^k$, $\mathbf{F}^k$ is inevitable to depend on the cutting location. FTM is applied again in order to obtain a feature of invariant cutting location. With FTM, the effect caused by the cutting location where the ring is cut can be eliminated. The learned local feature is constructed by concatenating all the $\mathbf{F}^k$ into a $K \times N/P$-dimensional vector.

According to the structure of CCRBM, the energy function of CCRBM and the joint probability distribution over the virtual layer $\mathbf{V}$ and the detection layer $\mathbf{H}$ can be defined as (13) and (14), respectively. The response of i-th node in virtual layer $\mathbf{V}$ is denoted as v_i, and the response of the j-th node in detection layer $\mathbf{H}^k$ is denoted as h_j^k. All nodes v_i in $\mathbf{V}$ have the same bias c, and also, all nodes h_i^k share a same bias b^k.

$$E(\mathbf{v}, \mathbf{h}) = - \sum_{k=1}^{K} \sum_{j=1}^{N} \sum_{r=1}^{m} h_j^k W_r^k v_{(j-1)m+r} - \sum_{k=1}^{K} b_k \sum_{j=1}^{N} h_j^k - c \sum_{i=1}^{N \times m} v_i \tag{13}$$

$$P(\mathbf{v}, \mathbf{h}) = \frac{1}{Z} exp(-E(\mathbf{v}, \mathbf{h})) \tag{14}$$

where Z is also defined as $\sum_{\mathbf{v}} \sum_{\mathbf{h}} exp(-E(\mathbf{v}, \mathbf{h}))$ which is similar to (1).

From (13), the nodes in the detection layer are conditionally independent of one another given the virtual layer $\mathbf{V}$ and vice versa. The virtual layer of CCRBM is real-valued; the virtual nodes conditioned on the detection layer are Gaussian with diagonal covariance. Thus, block Gibbs sampling can be efficiently performed by alternatively sampling each node in virtual layer $\mathbf{V}$ given the detection layer $\mathbf{H}$ in parallel and vice versa.

Similar to RBM, the following conditional distributions (15) and (16) are employed to perform block Gibbs sampling. Given the virtual layer $\mathbf{V}$, the detection layer $\mathbf{H}^k$ is sampled with the probability defined in (15). Analogously, given the detection layer $\mathbf{H}^k$ ($k \in [1, K]$), the virtual layer $\mathbf{V}$ is sampled using the probability defined in (16).

$$P(h_j^k = 1|\mathbf{v}) = sigmoid((W^k * v)_j + b_k) \tag{15}$$

$$P(v_i = 1|\mathbf{h}) = sigmoid(\sum_k (W^k * h^k)_i + c) \tag{16}$$

In principle, the parameters of CCRBM can be optimized by performing stochastic gradient ascent on the log-likelihood of training data. However, computing the exact gradient of the log-likelihood is intractable. Thus, contrastive divergence [31] is also employed to approximate the gradient of the log-likelihood.

With the probabilistic max-pooling layer, CCRBM is able to shrink the detection layer into high-level representation. Accordingly, the energy function of probabilistic max-pooling CCRBM can be defined in (17) by adding a constraint to (13), where the constraint summarizes the max-pooling process. Specifically, the detection nodes in the pooling window B_p are connected with the corresponding pooling node in a single potential way. In other words, at most one of the detection nodes in the pooling window is on, and the corresponding pooling node is on if and only if a detection node is on.

$$\begin{aligned} E(\mathbf{v}, \mathbf{h}) = \\ -\sum_{k=1}^{K}(\sum_{j=1}^{N}\sum_{r=1}^{m} h_j^k W_r^k v_{(j-1)m+r} + b_k \sum_{j=1}^{N} h_j^k) - c \sum_{i=1}^{N\times m} v_i \\ s.t. \sum_{j\in B_p} h_j^k \le 1, \forall k, p \end{aligned} \quad (17)$$

Given the virtual layer $\mathbf{V}$, we can sample the detection layer $\mathbf{H}^k$ and the pooling layer $\mathbf{P}^k$ as follows. For the k-th filter, the feature extracted from $\mathbf{V}$ can be defined in (18).

$$I(h_j^k) \triangleq (W^k * v)_j + b_k \quad (18)$$

Considering a node h_j^k in the pooling window B_p, the energy change of $E(\mathbf{v}, \mathbf{h})$ caused by turning it on is $-I(h_j^k)$ according to (18) and (13). The conditional probability used by Gibbs sampling is defined in (19) and (20).

$$P(h_j^k = 1|\mathbf{v}) = \frac{exp(I(h_j^k))}{1 + \sum_{j'\in B_p} exp(I(h_{j'}^k))} \quad (19)$$

$$P(p_p^k = 0|\mathbf{v}) = \frac{1}{1 + \sum_{j'\in B_p} exp(I(h_{j'}^k))} \quad (20)$$

Thus far, the procedure of sampling detection layer $\mathbf{H}^k$ conditioned on the virtual layer $\mathbf{V}$ is explained. The reverse procedure, sampling the virtual layer $\mathbf{V}$ given the detection layer $\mathbf{H}$, can be performed by Gibbs sampling using the conditional probability defined in (16).

3.6 Circle Convolutional DBN (CCDBN)

CCDBN is a hierarchical generative model for 3D local regions. Similar to CDBN, CCDBN consists of several max-pooling CCRBMs which are stacked layer-by-layer. The network of CCDBN represents an energy function by accumulating the energy functions for all stacked CCRBMs. The training of CCDBN also follows the greedy and layer-wise way. In short, once a lower level of CCRBM is trained, its weights are fixed, and its activations are used as the input to the next level of CCRBM.

Although the max-pooling CCRBM can be stacked into a CCDBN, we only test the max-pooling CCRBM in all experiments with the following reasons. Since PDD is a distribution, noise can be introduced into the PDD if the resolution of PDD is too fine (i.e., high number of bins), leading to degeneration of discriminability. Thus, the PDD with coarse resolution (i.e., small number of bins) is usually preferred. However, if the PDD resolution is coarse, then the PDD would become insensitive to small geometric changes. More specifically, the geometric changes captured by PDD are caused by the rotation of the sector window. If a sector window rotates with a small stride angle and the number of bins in PDD is small, the changes may become insignificant and even ignored. To alleviate this issue, the sector window is usually specified with a sufficiently large stride angle so that the geometric changes can be highlighted even the PDD resolution is coarse.

Nevertheless, it must be noted that a sufficiently large stride angle will sample fewer nodes from the pooling layer to the next level of CCRBM, leading to reduced amount of information. As a trade-off, a stride angle of 10° is employed in the experiments to capture structural changes between two adjacent sector windows.

4 Experimental Setup

In this section, different shape benchmark and performance measures for 3D local feature learning are, respectively, described for the following common areas of 3D shape analysis: global shape retrieval, partial shape retrieval, and shape correspondence. In addition, the setup of parameters for CCRBM is discussed.

4.1 Global Shape Retrieval

The objective of global shape retrieval is to search its globally similar shapes in a large dataset given a query shape. For the global shape retrieval experiment, the parameters of a CCRBM are firstly investigated using McGill 3D shape benchmark [38]. Then, the local features learned by CCRBM are compared with the ones obtained by other state-of-the-art feature descriptors under the articulated subset of McGill 3D shape benchmark, the whole set of McGill 3D shape benchmark, and SHREC 2007 dataset [39], respectively.

The McGill 3D shape benchmark contains 19 classes of shapes with totally 457 objects, which consists of an articulated subset and a non-articulated subset. The articulated subset contains 255 objects of 10 shape classes. SHREC 2007 dataset contains 400 shapes evenly distributed into 20 shape classes.

The statistical analysis of learned local features is performed to characterize a whole 3D shape using the Bag-of-Words (BoW) framework. The similarity measure $S(M_i, M_j)$ between two arbitrary shapes M_i and M_j is defined as the Chi-squared distance between their BoW feature vectors BoW_i and BoW_j as shown in (21).

$$S(M_i, M_j) = \frac{1}{2} \sum_{q} \frac{(BoW_i(q) - BoW_j(q))^2}{(BoW_i(q) + BoW_j(q))} \tag{21}$$

For all experiments, the traditional precision and recall (PR) curve is employed as the performance measure of the learned local features. Precision is the ratio of retrieved objects that are relevant to all retrieved objects in the ranked list. Recall is the ratio of relevant objects retrieved in the ranked list to all relevant objects in the dataset.

4.2 Partial Shape Retrieval

Partial shape retrieval is to efficiently search partially similar shapes with a query in a database. SHREC07 partial retrieval dataset is employed to evaluate the current experiment. Given the database of 400 watertight models and a set of 30 hybrid query models, each query is associated with a set of highly and marginally relevant items. In other words, these two sets of items represent, respectively, the classes which share a similar subpart with the query and those ones which are reasonably similar to the query. The quantitative evaluation is computed using the normalized discounted cumulated gain vector (NDCG) [40] via normalizing discounted cumulated gain vector (DCG) by the ideal cumulated gain vector calculated from the ground truth. DCG is defined in (22).

$$DCG[i] = \begin{cases} G[i] & \text{if } i = 1; \\ DCG[i-1] + G[i]/\log_2 i & \text{otherwise} \end{cases} \tag{22}$$

where $G[i]$ is a gain value representing the relevance of the i-th retrieved model (2 for highly relevant, 1 for marginally relevant, and 0 otherwise). For a given query, $DCG[i]$ is first computed, and then, $NDCG[i]$ is obtained by dividing the $DCG[i]$ by the ideal cumulated gain vector calculated from the ground truth.

4.3 Shape Correspondence

Shape correspondence aims to establish a meaningful relation between two sets of points on two given shapes. The features learned by CCRBM are also evaluated on shape correspondence over two datasets with well-defined ground truth. One dataset is SCAPE [41] which contains 71 mesh models representing a human body in different poses. Another dataset, Watertight dataset [42], contains 400 mesh models which are evenly classified into 20 object categories including humans, octopus, four-legged animals, ants, etc.

Given two shapes M_1 and M_2, the error measure proposed in the literature [42] is adopted to evaluate the accuracy of a predicted shape correspondence $f : M_1 \rightarrow M_2$ with respect to the ground truth map f_{true}. For every point b on shape M_1, the error between the predicted corresponding point $f(b)$ (on M_2) and the ground truth corresponding point $f_{\text{true}}(b)$ (on M_2) is measured by the distance of the shortest path connecting $f(b)$ to $f_{\text{true}}(b)$ on the M_2 surface (geodesic distance between $f(b)$ and $f_{\text{true}}(b)$ on M_2), which is denoted as $d_{M_2}(f(b), f_{\text{true}}(b))$. The smaller $d_{M_2}(f(b), f_{\text{true}}(b))$ is, the closer the estimated corresponding point to the ground truth. To eliminate the effect of scale of shapes, the $d_{M_2}(f(b), f_{\text{true}}(b))$ is normalized by the longest geodesic distance between any two points on the surface of M_2. Then, the accuracy between f and f_{true} is depicted by the probability distribution function of $d_{M_2}(f(b), f_{\text{true}}(b))$ of all points b on M_1.

4.4 The Setup of Parameters for CCRBM

Since global features can be constructed with the learned local features using the simple Bag-of-Words (BoW) framework, the evaluation of the BoW global feature can reflect the discriminability of the learned local features. Therefore, the parameters governing the learned local features of CCRBM are investigated over global shape retrieval task only.

CCRBM takes a set of five parameters including the central angle θ_c and the stride angle θ_s (which specify a sector window), the radius of the local region R, the number of filters K, and the number of training samples T. As discussed in the last paragraph in Sect. 3.6, the stride angle θ_s is carefully adjusted through experiments (in our case, $\theta_s = 10°$) in order to preserve the geometric and structural changes between two adjacent steps of a sector window as much as possible. Accordingly, the number of nodes N in the detection layer $\mathbf{H}^k$ is 36 ($360°/10°$). Moreover, the length of the pooling window P is set to 4, since a big width of the pooling window will shrink too much information from the detection layer $\mathbf{H}^k$, while a small width of the pooling window will make no sense about learning geometric and structural information from progressively larger input regions. In current experiments, $P = 4$ works very well. Note that θ_s and P are kept fixed in all experiments.

Since BoW framework is employed to combine the learned features as the global shape feature of an object, the number of virtual words K_w is another parameter to be determined. The framework and its component used in the experiments are introduced as follows:

The BoW framework The key idea of BoW is to represent a set of features as a sparse vector of occurrence counts of virtual words which come from a constructed vocabulary (so-called codebook). Each feature in a set is assigned to the closest virtual word according to the similarity between the feature and the virtual word, which provides an occurrence count to the closest virtual word when constructing a BoW vector.

Codebook construction To construct a codebook, a set of features are clustered into K_w clusters. Each of the cluster centroids serves as a virtual word in a codebook. Note that all features are learned by CCRBM from local regions centered at sampled vertices on each shape in the dataset. In our experiments, we sample 500 vertices on each 3D shape using farthest geodesic sampling method [34] as shown in Fig. 5a.

Specifically, the learned feature of each sampled vertex i from the j-th shape M_j is denoted as M_j^i, and then, k-means is used to cluster all features M_j^i into K_w clusters. The K_w centroids are regarded as the virtual words in a codebook $D = \{\bar{D}_1, \bar{D}_2, \ldots, \bar{D}_{K_w}\}$. Then, every vertex sampled on a shape M_j is associated with the closest visual word in D so that the shape M_j can be represented by a sparse vector BoW_j of occurrence counts of the visual words over all sampled vertices on it.

The number of virtual words K_w Two experiments were carried out to explore how the number of virtual words K_w affects performance on shape retrieval.

In both experiments, CCRBM was provided with a training set composed of 10% ($T = 10\%$) randomly selected sampled vertices from each shape. The CCRBM used $K = 300$ filters to detect each local region, each of which was centered at a selected vertex with a radius $R = 0.07\,\text{LGD}$ (the longest geodesic distance of that shape). The difference between the two experiments lies in the center angle θ_c of the sector window, which is set to 30° and 100°, respectively, for the two experiments.

By varying K_w, the performance of shape retrieval was evaluated through precision and recall (PR) curves. In the first experiment ($\theta_c = 30°$), PR curves were compared under $K_w = 10, 20, 30, 40, 50, 100$, and 200 visual words as shown in Fig. 8a. In the second experiment ($\theta_c = 100°$), PR curves were obtained under $K_w = 30, 40, 50, 60, 70, 80, 90, 100$, and 150 visual words as shown in Fig. 8b.

From Fig. 8, it is obvious that the retrieval performance obtained with 30, 40, and 50 visual words is fairly very well in both experiments. Thus, as a trade-off between the discriminative ability of BoW and computation efficiency, 40 visual words are employed to construct the global shape feature BoW_j in the following shape retrieval experiment.

The numbers of training samples T and filters K Two experiments were conducted to investigate the number of training samples T and the number of filters K of CCRBM.

Firstly, the CCRBM used $K = 300$ filters to learn each local region with a radius of $R = 0.07\,\text{LGD}$, and $\theta_c = 30°$. PR curves are obtained by CCRBM trained with $T = 6\%, 10\%, 20\%$, and 30% of all sampled vertices, respectively, as shown in Fig. 9a. It can be found that better retrieval performance can be obtained with

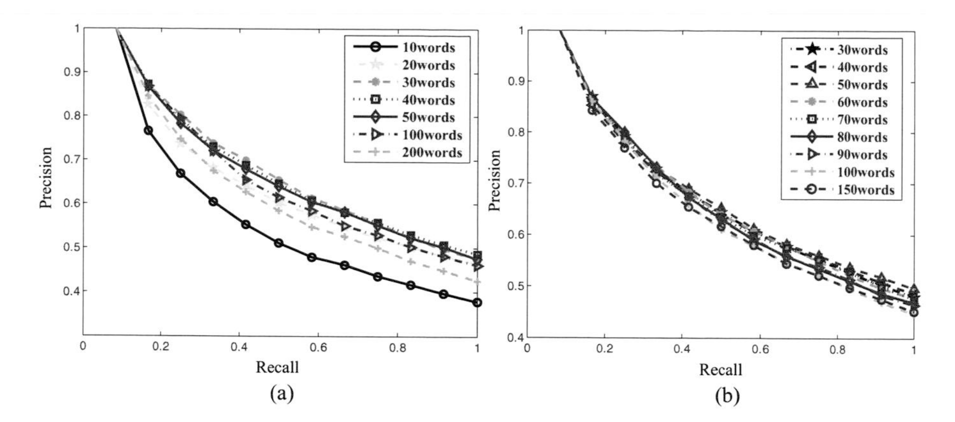

Fig. 8 The PR curves obtained under different numbers of virtual words (**a**) with $\theta_c = 30°$. (**b**) $\theta_c = 100°$

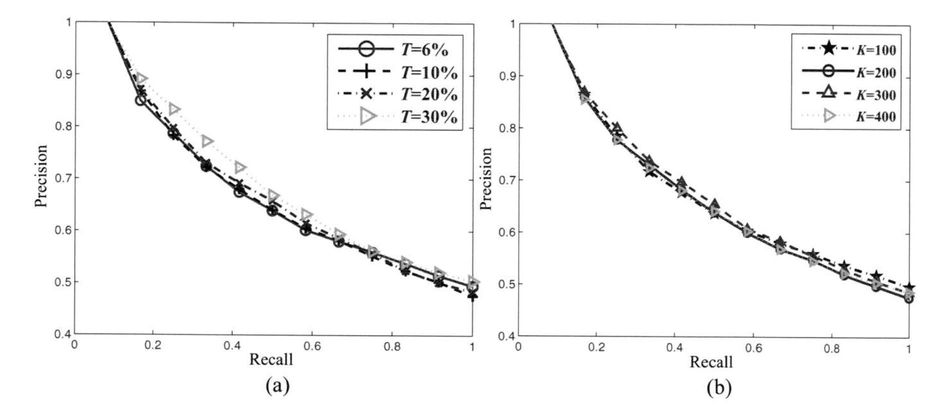

Fig. 9 The PR curves obtained with (**a**) different numbers of training samples T and (**b**) different numbers filters K

more training samples due to the acquisition of richer geometric and structural information of the local region, which abstracts high-level features.

Secondly, $T = 30\%$ sampled vertices were tried to train CCRBM with varying number of filters, while other parameters are the same as the previous experiments. The PR curves are illustrated in Fig. 9b, and the highest performance can be obtained with $K = 300$ filters. In the subsequent experiments, $K = 300$ filters are fixed to learn each local region.

The central angle θ_c and the radius of the local region R To determine θ_c, a CCRBM with K = 300 filters using $T = 10\%$ sampled vertices was tested, where each sampled vertex contains a neighbor of $R = 0.11$ LGD. PR curves with $\theta_c = 30°, 50°, 100°$, and $150°$ are obtained and shown in Fig. 10a. From the PR curves, the retrieval performance of CCRBM is affected little by θ_c. The reason is that most of the local regions can be modeled by symmetrical primitives, such as plane, cone, hemisphere, etc. Hence, the PDDs captured in sector windows with different central angles are nearly identical.

For setup of R, CCRBMs with a fixed sector window of $\theta_c = 50°$ were constructed with varying R, while the other parameters were the same as previous experiments. R took the values of 0.07, 0.09, 0.11, and 0.13 times LGD, respectively. Figure 10b shows that the PR curve with $R = 0.11$ LGD is the best because small local regions ($R = 0.07, 0.08$ LGD) look similar with each other which provide little amount of information for PDD to distinguish themselves into some basic primitive shapes. On the contrary, a large local region ($R = 0.13$ LGD) may contain highly irregular and overcomplex structure to capture distinguishable PDD information. As a result, the local regions with $R = 0.11$ LGD are selected for the training of CCRBM.

CCRBM parameters In summary, a CCRBM trained from McGill dataset using the following parameters as listed in Table 1 is involved in the experiments below.

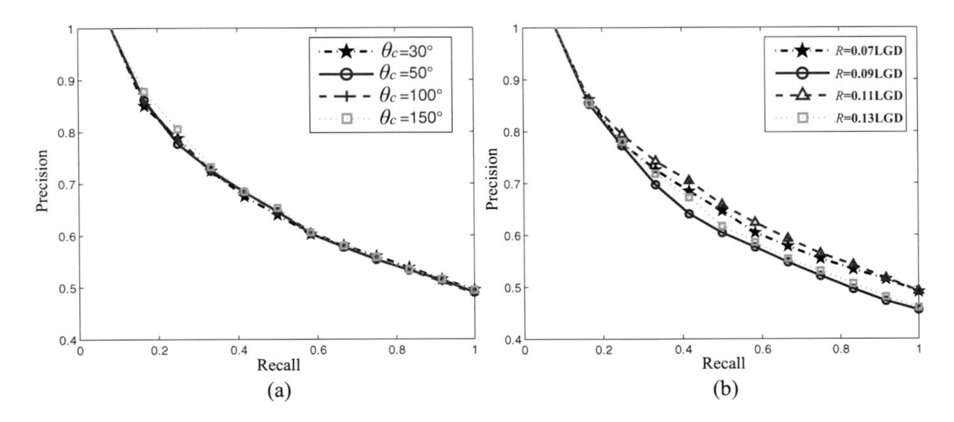

Fig. 10 The PR curves obtained with (**a**) different central angle of the sector window θ_c and (**b**) with different radius of local region R = a ratio times the longest geodesic distance (LGD)

Table 1 The parameters used in the following experiments

Virtual words K_w	Region size R	Filter number K	Training sample numbers T	Central angle θ_c	Stride angle θ_s	Pooling window size P
40	0.11 LGD	300	30%	50°	10°	4 nodes

5 Results and Analysis

5.1 Global Shape Retrieval

The retrieval performance was evaluated on the BoW global features which were constructed with local features learned by CCRBM. The virtual vocabulary was constructed using reduced features of 700 dimensions acquired from the originally learned local features of 2700 dimensions ($K \times N/P$) by principal component analysis (PCA). Through this reduction, a better retrieval performance can be obtained since the BoW global features cannot be constructed very well using high-dimensional features. The retrieval performance of CCRBM against other feature descriptors was evaluated over three datasets: (i) articulated subset of McGill 3D shape benchmark, (ii) whole McGill 3D shape benchmark, and (iii) SHREC2007 shape dataset.

Over the articulated subset of McGill 3D shape benchmark, the BoW global feature generated from CCRBM was compared to the moment descriptor (Moment) [43], shapeDNA [44], heat kernel signature (HKS) [23], spin image (SI) [10], and intrinsic spin image (ISI) [45]. The PR curves obtained by these methods are shown in Fig. 11a, in which the proposed CCRBM obtains the best result among all methods.

CCRBM was further compared with light field descriptor (LFD) [46], spherical harmonic descriptor (SHD) [47], and 3D ShapeNet on the whole McGill 3D shape benchmark. The PR curves obtained by these methods are portrayed in Fig. 11b. Once again, the results show that the features learned using CCRBM achieve the best retrieval result over other three state-of-the-art features. To obtain the results of 3D ShapeNet, we use the same parameters and code provided from the author [27]. As analyzed before, 3D ShapeNet cannot resist the rigid and nonrigid transformation of 3D shapes, and thus, the obtained result is not satisfactory.

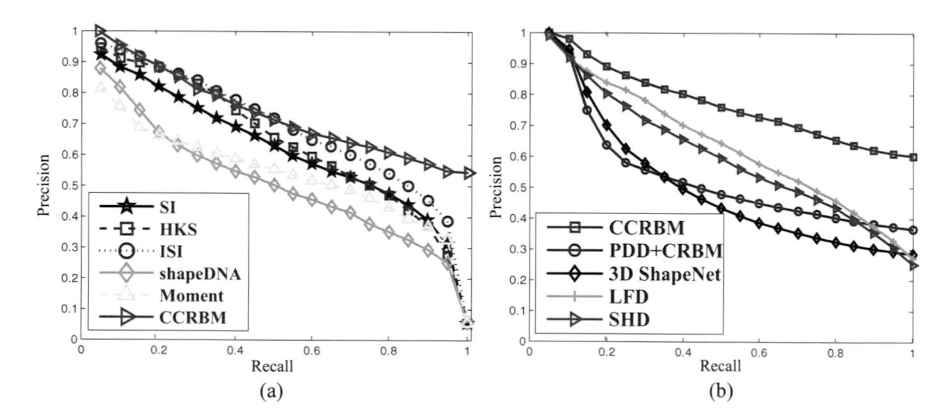

Fig. 11 PR curves on (**a**) the articulated subset of McGill shape benchmark; (**b**) the whole Mcgill shape benchmark

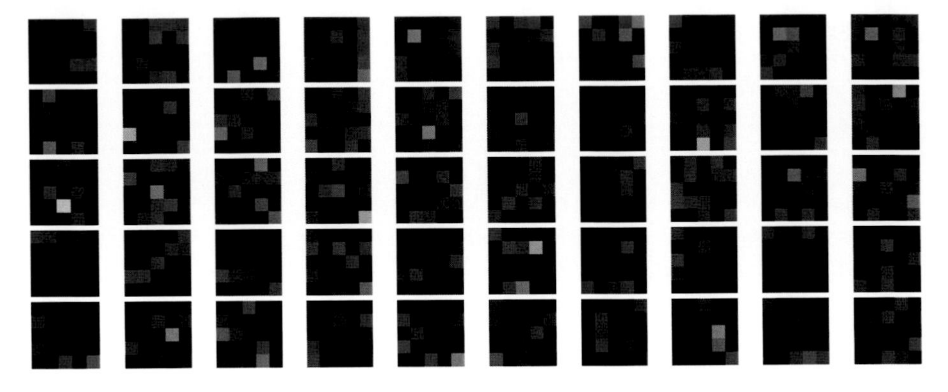

Fig. 12 The learned filters are shown briefly. Each 25-dimensional filter is shown as a 5 × 5 image

Lastly, the SHREC 2007 dataset was employed to evaluate CCRBM in global shape retrieval task. The comparison was against four state-of-the-art shape descriptors including the Hybrid BoW of Lavoue [48], the curve-based method of Tabia [49], the BoW method of Toldo et al. [50], covariance descriptors of Tabia [51], and 3D ShapeNet. The results are shown in Fig. 13a. CCRBM obtains better results than those by Tabia [51], Toldo [50], Lavoue [48], and 3D ShapeNet. Moreover, the result of CCRBM is comparable to that of Tabia [51]. Note that the CCRBM in this experiment was trained using the McGill shape benchmark, in which the training shapes were completely irrelevant to the test shapes in the SHREC 2007 dataset. Hence, the excellent performance of CCRBM implies that CCRBM is able to learn a highly general representation of 3D local region.

In addition, to highlight the strength of the ring-like structure and circle convolution of CCRBM, we compare results of CRBM and CCRBM under the McGill dataset in Fig. 11b and SHREC 2007 dataset in Fig. 13a, respectively. The result obtained by CRBM is using the same parameters of CCRBM but regarding the PDD of each local region as an image. The comparison implies that the structure of local regions preserved by the ring-like structure and circle convolution is important to the discriminative ability of the learned features. The learned filters of CCRBM are shown in Fig. 12. For better visualization, each 25-dimensional filter is transformed into a 5 × 5 image.

5.2 Partial Shape Retrieval

CCRBM was tested on partial shape retrieval using the SHREC07 partial retrieval dataset. The similarity between two shapes is also defined in (21). Figure 13b shows the NDCG curves of CCRBM and other state-of-the-art methods, which include not only the four methods that we have introduced in the Fig. 13a but also the extended Reeb graphs (ERG) [52], the curve skeleton-based many-to-many

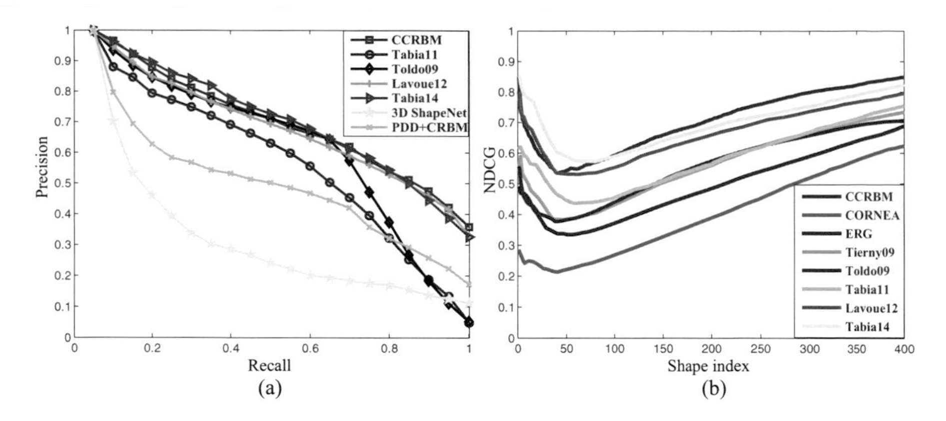

Fig. 13 PR curves of CCRBM compared to recent state-of-the-art descriptors under (**a**) the SHREC 2007 global shape retrieval dataset and (**b**) SHREC 2007 partial retrieval dataset

matching (CORNEA) [53], and the graph-based technique of Tierny et al [54]. It can be clearly seen that CCRBM outperforms all other methods in the experiment. The discriminability of the local features learned by CCRBM is benefited from both the use of PDD and the structure preserving learning diagram which captures more geometric and structural information.

5.3 Shape Correspondence

Different from global and partial shape retrieval focusing on statistical feature, we further investigate the effect of pairwise correspondence between two sample points from two different shapes. In this experiment, the sampled points from shape M_1 are matched to the sampled points from shape M_2. For all shapes in the dataset, two shapes are selected as a matching pair when these two shapes are assigned with the subsequent index. For example, shape 1 and shape 2 form the first matching pair and shape 2 and shape 3 form the second matching pair and so on. In this experiment, no complicated matching algorithm was employed but the simple Hungarian algorithm [55] to match sampled points in the feature space, which aimed to explore the raw discriminative power of the learned features. Here, the Hungarian algorithm was used as a combinatorial optimization algorithm that establishes the correspondence between two sets of sampled points via minimizing the overall correspondence cost. The correspondence cost of each pair of points is the L2 norm of difference between their learned local features.

CCRBM was compared with the heat kernel signature (HKS) [23] on Watertight dataset and SCAPE dataset, respectively. Figure 14 depicts the curves that portray the probability distribution function of the errors, which is the ratio of correct matches vs ratio of geodesic error curves. The x-axis of the curve represents

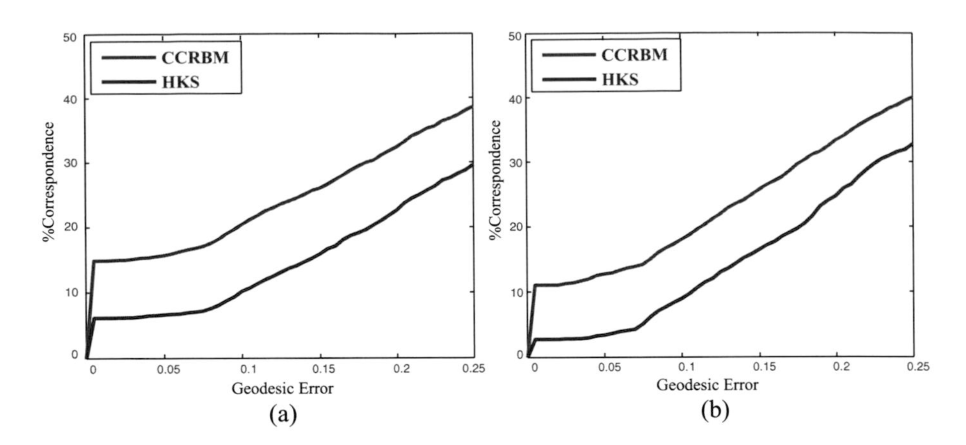

Fig. 14 Ratio of correct matches vs ratio of geodesic error curves obtained by HKS and the features learned by CCRBM under (**a**) Watertight dataset and (**b**) SCAPE dataset

a varying geodesic distance threshold, D_C, and the y-axis shows the average percentage of points for which $d_{M_2}(f(p), f_{\text{true}}(p)) < D_C$. The desired percentage of correct matches should be large when the error is fixed. It is clearly seen in Fig. 14a that CCRBM always lies above HKS on Watertight dataset. This phenomenon is consistently observed in Fig. 14b, which shows dramatically better performance when comparing our learned local features to the handcrafted HKS.

5.4 Significance and Analysis

CCRBM is significant in learning geometry and structure information from huge amount of 3D patches with resolving three challenging issues, the irregular vertex topology, orientation ambiguity on the surface, and the effect of rigid transformation. In fact, different training patches may be produced under different rigid transformations over finite patch patterns, such as plane, arc surface, cone surface, etc. By tackling these three issues, redundant information in training samples caused by different rigid transformations can be dramatically reduced so that CCRBM can effectively learn common feature patterns from the training samples. Simply speaking, in CCRBM, no matter how vertices are connected with each other to form a patch, no matter how a patch is translated or rotated, and where the starting location of the filters is, the learned geometry and structure information becomes invariant. This kind of learning manner is a promising characteristic for 3D surface that is missed in the existing deep learning models. That explains why CCRBM works better than other existing methods.

From the experiment results, it can be observed that the high performance of CCRBM stems from the innovative idea of circular convolution, PDD, FTM, and the ring-like model structure. These enable CCRBM to learn general and discriminative feature patterns. With the novel ring-like structure of CCRBM and the circle convolution, the geometry information of the local region is captured along with fully preserving the spatial structure. Capturing and encoding the geometry and structure information together help increase the power of discriminating different local regions. Furthermore, the novel PDD and FTM are the elements to guarantee the ability of learning general feature patterns. These two elements provide the invariant raw feature for CCRBM with eliminating the ambiguous initial location of the circular convolution window and resisting the scale, the rigid transformation, and the nonrigid transformation of 3D shapes.

5.5 Limitation and Future Work

Although the proposed CCRBM achieves a high performance on retrieval and correspondence, there are still two limitations. *One is that CCRBM cannot work on a highly noisy or non-manifold surface due to the learning of raw 3D feature based on PDD. The noise on the surface may significantly corrupt the true geometry and structure of local regions; in addition, the consistent size of local region cannot be precisely determined on the non-manifold surface.* Another limitation is that a sufficiently large stride angle is necessary to capture the geometric and structural changes between adjacent sector windows, which may however sample insufficient number of nodes in the pooling layer to be conveyed to next level of CCRBM.

In the future, the generality of CCRBM can be further explored and analyzed. An interesting idea comes up that various basic primitive shapes with clean surface, such as square, cylinder, sphere, cone, etc., can be employed to train a CCRBM, and the trained CCRBM can try to extract more discriminative features from arbitrary 3D shapes for different tasks in 3D shape analysis. *In addition, we will try to employ the depth images to help CCRBM learn real isolated 3D shapes. Although single view depth images are always incomplete because of the field of view and occlusion, which makes it hard to determine continuous and consistent size of local regions and calculate reliable PDD over local regions, we plan to use CCRBM to learn real 3D shapes reconstructed from multi-view depth images.*

6 Conclusion

CCRBM for unsupervised 3D local feature learning is proposed, which aims to overcome the disadvantages of handcrafted descriptors and the difficulties of learning 3D data directly via deep learning models. CCRBM significantly resolves three challenging issues, namely, the irregular vertex topology, the effect of rigid transformations, and the orientation ambiguity on the surface that the existing deep learning models cannot resolve. These significant merits come from the combination of the novel elements of CCRBM, including PDD, circle convolution, FTM, and the ring-like model structure.

Comparison of CCRBM with other state-of-the-art feature descriptors including light field descriptor (LFD), spherical harmonic descriptor (SHD), the moment descriptor (Moment), shapeDNA, heat kernel signature (HKS), spin image, intrinsic spin image, covariance descriptors, etc. has been conducted on various experiments evaluated on three common works of shape analysis: global shape retrieval, partial shape retrieval, and shape correspondence. Experimental results show that the CCRBM learns more discriminative features and achieves higher retrieval and correspondence accuracy than other feature descriptors. Finally, the significant contributions of our work are summarized as follows:

1. A novel deep learning model called CCRBM for unsupervised 3D local feature learning is proposed.
2. A novel convolution manner, circle convolution, is proposed to perform the convolution operation on 3D shape surfaces directly along a fixed orientation. In addition, FTM is introduced to eliminate the ambiguous starting location of the convolution.
3. The novel ring-like multilayer structure of CCRBM is specifically designed to capture the geometric and structural variations of 3D local regions along a circle direction sequentially. The novel model structure of CCRBM is highly suitable for 3D data and much different from the ones of existing deep learning models.
4. PDD is innovatively proposed as a raw 3D representation which is only determined by coordinates of vertices in a 3D local region.
5. A deep architecture CCDBN is proposed as a stacked multilayer CCRBMs.

References

1. Y. Gao, M. Wang, D. Tao, R. Ji, and Q. Dai, 3D object retrieval and recognition with hypergraph analysis, IEEE Transactions on Image Processing, vol. 21, no. 9, pp. 4290–4303, 2012.
2. O. van Kaick, H. Zhang, G. Hamarneh, and D. Cohen-Or, A survey on shape correspondence, Computer Graphics Forum, vol. 30, no. 6, pp. 1681–1707, Sep 2011.
3. X. Qian and C. Ye, Ncc-ransac: A fast plane extraction method for 3D range data segmentation, IEEE Transactions on Cybernetics, vol. 44, pp. 2771–2783, 2014.
4. K.-C. Chan, C.-K. Koh, and C. S. G. Lee, A 3-d-point-cloud system for human-pose estimation, IEEE Transactions on Cybernetics, vol. 44, pp. 1486–1497, 2014.
5. O. S. Gedik and A. A. Alatan, 3D rigid body tracking using vision and depth sensors, IEEE Transaction on Cybernetics, vol. 43, no. 5, pp. 1395–1405, 2013.
6. M. Liang, H. Min, R. Luo, and J. Zhu, Simultaneous recognition and modeling for learning 3D object models from everyday scenes, IEEE Transactions on Cybernetics, vol. PP, pp. 1–12, 2014.
7. F.-F. Li and P. Perona, A bayesian hierarchical model for learning natural scene categories, Proceedings of IEEE Conference on Computer Vision and Pattern Recognition, vol. 2, 2005, pp. 524–531.

8. A. M. Bronstein, M. M. Bronstein, L. J. Guibas, and M. Ovsjanikov, Shape google: geometric words and expressions for invariant shape retrieval, ACM Transactions on Graphics, vol. 30, no. 1, pp. 1–20, 2011.
9. L. Shapira, A. Shamir, and D. Cohen-Or, Consistent mesh partitioning and skeletonisation using the shape diameter function, The Visual Computer, vol. 24, no. 4, pp. 249–259, 2008.
10. A. Johnson and M. Hebert, Using spin images for efficient object recognition in cluttered 3D scenes, IEEE Transactions on Pattern Analysis and Machine Intelligence, vol. 21, no. 5, pp. 433–449, 1999.
11. T. Darom and Y. Keller, Scale-invariant features for 3D mesh models. IEEE Transactions on Image Processing, vol. 21, no. 5, pp. 2758–2769, 2012.
12. Y. Bengio, A. Courville, and P. Vincent, Representation learning: A review and new perspectives, IEEE Transactions on Pattern Analysis and Machine Intelligence, vol. 35, no. 8, pp. 1798–1828, 2013.
13. H. Yan, J. Lu, and X. Zhou, Prototype-based discriminative feature learning for kinship verification, IEEE Transactions on Cybernetics, vol. PP, pp. 1–13, 2014.
14. X. Lu, Y. Yuan, and P. Yan, Alternatively constrained dictionary learning for image superresolution, IEEE Transactions on Cybernetics, vol. 44, pp. 366–377, 2014.
15. H. Lee, R. Grosse, R. Ranganath, and A. Y. Ng, Convolutional deep belief networks for scalable unsupervised learning of hierarchical representations, in The Annual International Conference on Machine Learning, 2009, pp. 609–616.
16. A. Krizhevsky, I. Sutskever, and G. E. Hinton, Imagenet classification with deep convolutional neural networks, in Advances in Neural Information Processing Systems, 2012, vol. 25, pp. 1097–1105.
17. A. Graves, A. rahman Mohamed, and G. E. Hinton, Speech recognition with deep recurrent neural networks, in IEEE International Conference on Acoustics, Speech, and Signal Processing, 2013, pp. 6645–6649.
18. S. Bu, Z. Liu, J. Han, J. Wu, and R. Ji, Learning high-level feature by deep belief networks for 3D model retrieval and recognition, IEEE Transactions on Multimedia, vol. 16, no. 8, pp. 2154–2167, 2014.
19. Z. Liu, S. Chen, S. Bu, and K. Li, High-level semantic feature for 3D shape based on deep belief networks, Proceeding of IEEE International Conference on Multimedia and Expo, 2014, pp. 1–6.
20. B. Leng, X. Zhang, M. Yao, and Z. Xiong, A 3D model recognition mechanism based on deep boltzmann machines, Neurocomputing, vol. 151, pp. 593–602, 2014.
21. P. Heider, A. Pierre-Pierre, R. Li, and C. Grimm, Local shape descriptors, a survey and evaluation, in Eurographics Workshop on 3D Object Retrieval, 2011, pp. 49–57.
22. J. Lodder, Curvature in the calculus curriculum. The American Mathematical Monthly, vol. 110, no. 7, pp. 593–605, 2003.
23. J. Sun, M. Ovsjanikov, and L. J. Guibas, A concise and provably informative multi-scale signature based on heat diffusion, Computer Graphics Forum, vol. 28, no. 5, pp. 1383–1392, 2009.
24. M. M. Bronstein and I. Kokkinos, Scale-invariant heat kernel signatures for non-rigid shape recognition, Proceedings of IEEE Conference on Computer Vision and Pattern Recognition, 2010, pp. 1704–1711.
25. M. Aubry, U. Schlickewei, and D. Cremers, The wave kernel signature: A quantum mechanical approach to shape analysis, in International Conference on Computer Vision Workshops, 2011, pp. 1626–1633.
26. R. Socher, B. Huval, B. Bhat, C. D. Manning, and A. Y. Ng, Convolutional-recursive deep learning for 3D object classification, in Advances in Neural Information Processing Systems, 2012, vol. 25, pp. 665–673.
27. Z. Wu, S. Song, A. Khosla, X. Tang, and J. Xiao, 3D ShapeNets for 2.5D object recognition and Next-Best-View prediction, arXiv: 1406.5670, 2014.
28. G. E. Hinton, S. Osindero, and Y.-W. Teh, A fast learning algorithm for deep belief nets, Neural Computation, vol. 18, no. 7, pp. 1527–1554, 2006.

29. G. E. Hinton and R. R. Salakhutdinov, Reducing the dimensionality of data with neural networks, Science, vol. 313, pp. 504–507, 2006.
30. H. Ackley, E. Hinton, and J. Sejnowski, A learning algorithm for boltzmann machines, Cognitive Science, pp. 147–169, 1985.
31. G. E. Hinton, Training products of experts by minimizing contrastive divergence, Neural Computation, vol. 14, no. 8, pp. 1771–1800, 2002.
32. G. Casella and E. I. George, Explaining the gibbs sampler, The American Statistician, vol. 46, no. 3, pp. 167–174, 1992.
33. R. Osada, T. Funkhouser, B. Chazelle, and D. Dobkin, Shape distributions, ACM Transactions on Graphics, vol. 21, no. 4, pp. 807–832, 2002.
34. G. Peyre and L. D. Cohen, Geodesic remeshing using front propagation, International Journal of Computer Vision, vol. 69, no. 1, pp. 145–156, 2006.
35. S. Zokai and G. Wolberg, Image registration using log-polar mappings for recovery of large-scale similarity and projective transformations, vol. 14, pp. 1422–1434, 2005.
36. I. Kokkinos and A. L. Yuille, Scale invariance without scale selection, Proceedings of IEEE Conference on Computer Vision and Pattern Recognition, 2008, pp. 1–8.
37. I. Kokkinos, M. M. Bronstein, R. Litman, and A. M. Bronstein, Intrinsic shape context descriptors for deformable shapes, Proceedings of IEEE Conference on Computer Vision and Pattern Recognition, 2012, pp. 159–166.
38. K. Siddiqi, J. Zhang, D. Macrini, A. Shokoufandeh, S. Bouix, and S. Dickinson, Retrieving articulated 3D models using medial surfaces, Machine Vision and Applications, vol. 19, no. 4, pp. 261–275, 2008.
39. D. Giorgi, S. Biasotti, and L. Paraboschi, Shape retrieval contest 2007: Watertight models track, 2007.
40. S. Marini, L. Paraboschi, and S. Biasotti, Shape retrieval contest 2007: Partial matching track, in SHREC in conjunction with IEEE Shape Modelling International, 2007, pp. 13–16.
41. D. Anguelov, P. Srinivasan, H.-C. Pang, D. Koller, S. Thrun, and J. Davis, The correlated correspondence algorithm for unsupervised registration of nonrigid surfaces, Proceeding of the Neural Information Processing Systems, 2004, pp. 33–40.
42. V. G. Kim, Y. Lipman, and T. Funkhouser, Blended intrinsic maps, ACM Transaction on Graphics, vol. 30, no. 4, pp. 79:1–79:12, 2011.
43. A. Elad and R. Kimmel, On bending invariant signatures for surfaces, IEEE Transactions on Pattern Analysis and Machine Intelligence, vol. 25, no. 10, pp. 1285–1295, 2003.
44. R. Martin, W. F. Erich, and P. Niklas, Laplace-spectra as fingerprints for shape matching, in Symposium on Solid and Physical Modeling, 2005, pp. 101–106.
45. X. Wang, Y. Liu, and H. Zha, Intrinsic spin images: A subspace decomposition approach to understanding 3D deformable shapes, in 5th International Symposium 3D Data Processing, Visualization and Transmission, 2010, pp. 17–20.
46. D. Chen, X. Tian, Y. Shen, and M. Ouhyoung, On visual similarity based 3D model retrieval, Computer Graphics Forum, vol. 22, no. 3, pp. 223–232, 2003.
47. M. Kazhdan, T. Funkhouser, and S. Rusinkiewicz, Rotation invariant spherical harmonic representation of 3D shape descriptors, Proceedings of Eurographics Symposium on Geometry Processing, 2003, pp. 156–165.
48. G. Lavou, Combination of Bag-of-Words Descriptors for Robust Partial Shape Retrieval, The Visual Computer, vol. 28, no. 9, pp. 931–942, 2012.
49. H. Tabia, M. Daoudi, J.-P. Vandeborre, and O. Colot, A new 3D-matching method of non-rigid and partially similar models using curve analysis, IEEE Transactions on Pattern Analysis and Machine Intelligence, vol. 33, no. 4, pp. 852–858, 2011.
50. R. Toldo, U. Castellani, and A. Fusiello, Visual vocabulary signature for 3D object retrieval and partial matching. in Eurographics Workshop on 3D Object Retrieval, 2009, pp. 21–28.
51. H. Tabia, H. Laga, D. Picard, and P.-H. Gosselin, Covariance descriptors for 3D shape matching and retrieval, Proceedings of IEEE Conference on Computer Vision and Pattern Recognition, pp. 4815–4192, 2014.

52. S. Biasotti, S. Marini, M. Spagnuolo, and B. Falcidieno, Sub-part correspondence by structural descriptors of 3D shapes. Computer-Aided Design, vol. 38, pp. 1002–1019, 2006.
53. N. D. Cornea, M. F. Demirci, D. Silver, A. Shokoufandeh, S. J. Dickinson, and P. B. Kantor, 3D object retrieval using many-to-many matching of curve skeletons, Proceedings of the International Conference on Shape Modeling and Applications, 2005, pp. 368–373.
54. J. Tierny, J.-P. Vandeborre, and M. Daoudi, Partial 3D shape retrieval by reeb pattern unfolding. Computer Graphics Forum, vol. 28, pp. 41–55, 2009.
55. H. W. Kuhn, The hungarian method for the assignment problem, Naval Research Logistics Quarterly, vol. 2, pp. 83–97, 1955.

Deep Learning-Based Descriptors for Object Instance Search

Jie Lin, Olivier Morère, Antoine Veillard, and Vijay Chandrasekhar

Abstract In the past 5 years, deep learning has achieved remarkable breakthroughs, mainly attributed to the success of convolutional neural networks (CNN) on vision applications like ImageNet classification. In this chapter, we are interested in deep learning-based descriptors for object instance search in images. Specifically, we propose to tackle some practical issues of existing CNN models, with a focus on resource-efficient yet effective deep descriptors extracted from CNN. (1) How to achieve compact image representations (e.g., hundreds of bits) from deep neural networks in an end-to-end manner? (2) How to encode scale/rotation invariances into deep CNN architecture? To address the issues, our approach has two novel contributions. First, we propose Restricted Boltzmann Machine with a novel batch-level regularization scheme specifically designed for the purpose of descriptor hashing (RBMH), which is able to match the performance of the uncompressed descriptor with tiny 32–256 bit hashes. Second, inspired from invariance theory, we propose Nested Invariance Pooling (NIP), a method for computing group-invariant transformations with feed-forward neural networks. We specifically incorporate scale, translation, and rotation invariances, but the scheme can be extended to any arbitrary sets of transformations. A thorough empirical evaluation with state of the art shows that the results obtained both with the NIP descriptors and the NIP+RBMH hashes are consistently outstanding across a wide range of datasets.

J. Lin (✉)
Institute for Infocomm Research, Singapore, Singapore
e-mail: lin-j@i2r.a-star.edu.sg

O. Morère · A. Veillard
Université Pierre et Marie Curie, Paris, France

V. Chandrasekhar
Institute for Infocomm Research and Nanyang Technological University, Singapore, Singapore
e-mail: vijay@i2r.a-star.edu.sg

X. Jiang et al. (eds.), *Deep Learning in Object Detection and Recognition*,
https://doi.org/10.1007/978-981-10-5152-4_8

1 Introduction

Object instance search (Fig. 1) is the problem of finding an object instance present in a query image from a database of images (aka particular object retrieval). State-of-the-art object instance search pipelines consist of two major steps: first, a subset of images similar to the query are retrieved from the database, and second, Geometric consistency checks (GCC) are applied to select the relevant images from the subset with high precision. The first step is based on comparison of global image descriptors: high-dimensional vectors with up to tens of thousands of dimensions representing the image data. The second step is computationally highly complex and can only be applied to hundreds or thousands of images in practical applications. More discriminative global descriptors result in relevant images being more highly ranked, resulting in fewer images that need to be compared pairwise with GCC.

As a result, better global descriptors are key to improving search performance and have been the object of much recent interest. Scale, rotation, and orientation changes between query and database objects and background clutter pose significant challenges for this problem. Furthermore, fast searches in large databases of millions or even billions of images require the global descriptors to be compressed into compact representations. This chapter will focus on how to achieve extremely compact global descriptor representations for large-scale object instance search.

Recent works [11] show that image descriptors extracted from CNN intermediate layers (e.g., convolutional layer) offer better search performance than traditional handcrafted features (e.g., SIFT [52] and Fisher vector [58]) on average, but

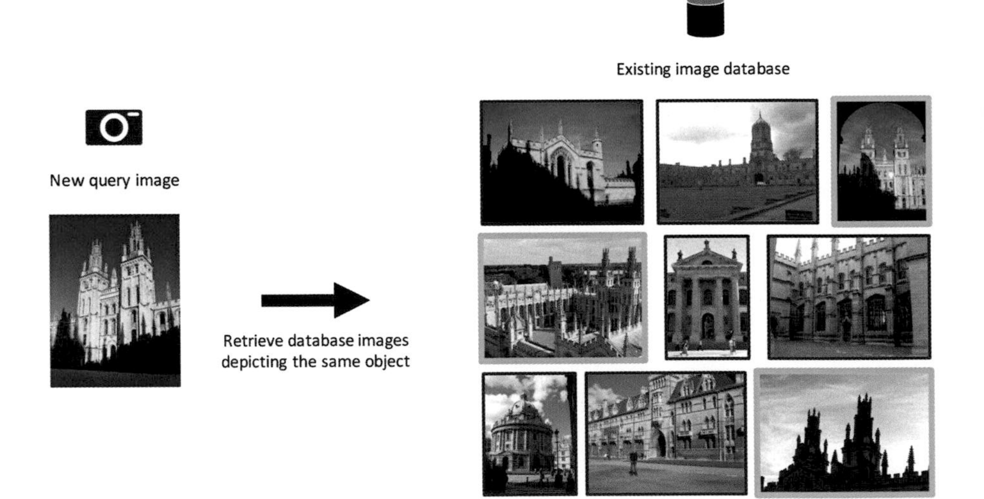

Fig. 1 In the object instance search problem, the task is to select database images depicting the same object instance as the one depicted in the query image. No external information is used (categories, labels...)

two issues related to CNN descriptors need to be addressed: first, the lack of transformation (specifically, scale and rotation) invariance of the CNN descriptors, and second, the high dimensionality and scalar nature of the descriptors making descriptor matching inefficient.

We tackle the first problem by hashing the descriptors to small binary codes for efficient matching with Hamming distances while retaining the good search performance of the uncompressed descriptors [12, 46–48]. The very low bitrate range of 32–1024 bits is specifically targeted. The proposed hashing pipeline first showed how high-dimensional descriptors can be compressed to very compact binary representations. Key to achieving excellent performance at low rates is the regularization of stacked Restricted Boltzmann Machines (RBM). Next is shown how Siamese fine-tuning can be used to improve performance. This technique is applicable where labeled external data of matching instance pairs is available.

To address the second problem, we propose Nested Invariance Pooling (NIP) [45, 50, 53, 54], a method to produce compact global image descriptors from visual representations extracted from CNN, which are robust to multiple types of image transformations like translations, rotations, and scale changes. NIP is inspired from *invariance theory* (i-theory) [3–5], a recently proposed mathematical theory for computing group-invariant transformations with feed-forward neural networks. We show that NIP produces compact (but non-binary) global image descriptors which outperform other schemes at equivalent descriptor dimensionality on most evaluation datasets. Finally, combined NIP and hashing pipeline produces some of the best performing hashes available in the literature, especially at very low bitrates (256 bits and lower).

2 Related Work

Handcrafted Global Descriptors State-of-the-art handcrafted global descriptors for object search are Fisher vector (FV) [58] and vector of locally aggregated descriptors (VLAD) [38], which are often aggregated from local descriptors such as SIFT [52] (Fig. 2). FV is obtained by quantizing the set of local feature descriptors with a small codebook of 64–512 centroids and aggregating first- and second-order residual statistics for features quantized to each centroid. The residual statistics from each centroid are concatenated together to obtain the high-dimensional global descriptor representation, typically 8192 to 65,536 dimensions. The performance increases as the dimensionality of the global descriptor increases, as shown in [58]. VLAD descriptor can be considered a special case of the FV, with hard quantization of feature descriptors, with concatenation of only first-order residual statistics in the final descriptor representation.

Several improvements have been proposed over the baseline VLAD and FV approaches. In [15], another normalization scheme is proposed for the residuals, where the per cluster mean of residuals is computed instead of the sum, enhancing the discriminativeness of the VLAD descriptor. In [38], a similar signed square

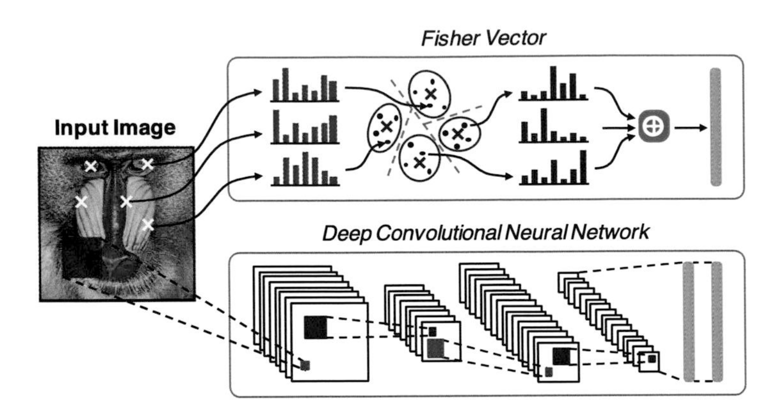

Fig. 2 Fisher vector (FV) and convolutional neural network (CNN)-based pipelines for the extraction of *global descriptors* from images. (Reprinted from Ref. [11], with permission from Elsevier)

rooting (SSR) normalization scheme is proposed for VLAD. In [36], VLAD is extended by using PCA whitening and multiple clusterings for quantization. Intra-normalization scheme is proposed in [6] to alleviate the adverse effect of bursty visual features [37], where the sum of residuals is L2 normalized within each VLAD block. FV is improved over the baseline approach of [58] by using nonlinear additive kernel and normalization. Other improvements to the baseline FV [58] include the residual enhanced visual vector [16], the scalable compressed Fisher vector [44], better matching kernels [65], and better aggregation schemes [39]. Some of the best reported instance search results are still based on handcrafted FV [65]. Next, the matter of how CNN have been applied for the object instance search problem is discussed.

Deep CNN Descriptors As opposed to the carefully handcrafted descriptors, there has been a fair bit of work on CNN descriptors for object instance search in recent literature [8, 9, 26, 60, 62, 66]. Razavian et al. [60] evaluate representations extracted from CNN fully connected layer on a wide range of tasks, including as a global descriptor for instance search, and show promising initial results. Then, Babenko et al. [9] show that a pre-trained CNN can be fine-tuned with domain-specific data (objects, scenes, etc.) to improve instance search performance on relevant datasets. In [8], Babenko et al. show how pooled intermediate layers of a CNN can be used as a starting representation for instance search. They show that sum-pooling of intermediate feature maps performs better than max-pooling, when the image representation is whitened. Note that the approach in [8] provides limited invariance to translation, but not to scale or rotation. In MOP-CNN [26], the authors propose extracting CNN activations using a sliding window approach at different scales in the image, followed by computing a high-dimensional VLAD representation on the local CNN descriptors. While this results in highly performant descriptors, the starting representations are often orders of magnitude larger than original descriptors. References [7, 62] show that spatial max-pooling of intermediate maps is an effective representation, and higher performance can be achieved compared to using the fully connected layers. Another very recent work [66] proposes pooling across regional bounding boxes in the image, similar to the popular R-CNN approach [21] used for object detection.

While deep learning has unquestionably become the dominant approach for many visual tasks, raw image descriptors from CNN do not systematically have the upper hand over FV in object instance search. The two types of descriptors being radically different in nature, one can expect them to behave very differently based on specific aspects of the problem. Unlike interest points which provide scale and rotation invariance to the FV pipeline, CNN representations used in image classification are obtained by densely sampling a resized canonical image. CNN do not have a built-in mechanism to ensure resilience to geometric transformations like scale and rotation and, therefore, do not provide explicit rotation and scale invariance, which are often key to instance search tasks. Moderate levels of scale and rotation invariance for CNN features are nevertheless indirectly achieved from the max-pooling operations in the pipeline, the diversity of the training data which typically contains objects at varying scales and orientations, and data augmentation during the training phase where data can be preprocessed and input to the CNN at different scales and orientations.

Hashing The dimensionality of such global descriptors is typically very high: 8192 to 65,536 floating point numbers for FV[58] and 4096 for CNN [41], while extremely compact image representations such as 64-bit hashes are a definite must for fast image instance search because (1) 64 bits provide more than enough capacity for any practical purposes, including Internet-scale problems and (2) a 64-bit hash is directly addressable in RAM and enables fast matching using Hamming distances. Bringing such high-dimensional representations down to a 64-bit hash is a considerable challenge.

While there is plenty of work on learning binary codes [29] for compressing small descriptors like SIFT, there is relatively little work on compression of high-dimensional global descriptors. Proposed methods for compression of descriptors like SIFT or GIST include [13, 14, 24, 29, 31, 42, 49, 56, 67, 69, 70]. The global descriptor data in consideration in this work are two orders of magnitude higher in dimensionality, making the problem significantly more challenging.

The most important hashing techniques applied to the global descriptor compression problem are reviewed here. Perronnin et al. [58] propose ternary quantization of FV, quantizing each dimension to $+1$, -1, or 0. The authors show that this representation results in little loss in performance. However, this results in descriptor size of thousands of bits. Perronnin et al. also explore locality-sensitive hashing [18] and spectral hashing [70]. Spectral hashing performs poorly at high rates, while LSH and simple ternary quantization need thousands of bits to achieve good performance. Gong et al. propose the popular iterative quantization (ITQ) scheme and apply it to the GIST descriptor in [24]. ITQ first performs principal component analysis (PCA) to reduce dimensionality and subsequently learns a rotation to minimize the quantization error of mapping the transformed data to the vertices of a zero-centered binary hypercube. One drawback of this scheme is that the PCA matrix might require several GBs of memory for high- dimensional global descriptors. In subsequent work, Gong et al. in [23] show how bilinear projections can be used to create binary hashes of VLAD [24]. Gong et al. [23] focus on generating very long codes for global descriptors, and the bilinear projection-based binary codes (BPBC) scheme requires tens of thousands of bits to match the performance of the uncompressed global descriptor.

The MPEG-CDVS standard adopted the scalable compressed Fisher vector [44], which was based on binarization of high-dimensional Fisher vector. The size of the compressed descriptor in the MPEG-CDVS standard ranges from 256 bytes to several thousand bytes per image, based on the operating point. Bit selection is performed greedily to maximize pairwise receiver operating characteristic (ROC) matching performance. Stacked Restricted Boltzmann Machines (RBM), primarily known as powerful dimensionality reduction techniques [61], can also be used for hashing.

3 Compact Invariant Deep Descriptors

In this section, we show overview of the proposed framework toward compact invariant deep representations for object instance search (in Fig. 3). With the output extracted from CNN, our method is composed of two key components: multigroup nested invariant pooling is proposed to derive translation, rotation, and scale-invariant representations (Sect. 3.3), followed by descriptor hashing scheme presented in Sects. 3.1 and 3.2, with the aim to achieve the best performing tiny 32–256 bits hashes.

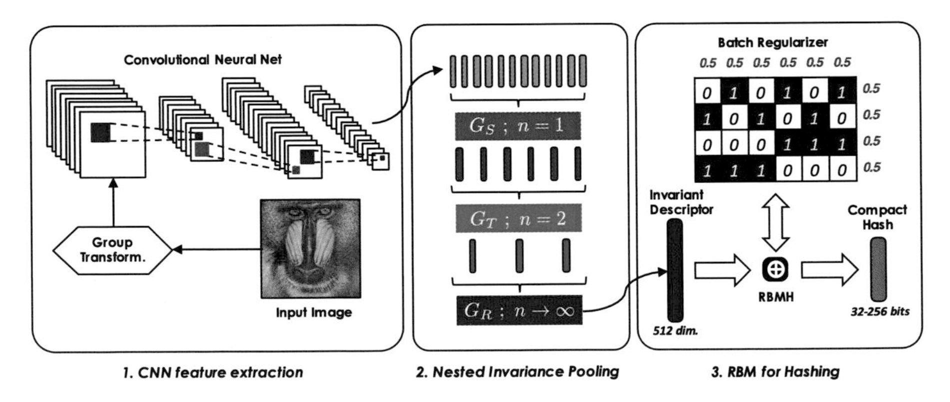

Fig. 3 Nested Invariance Pooling (NIP) to produce robust descriptors from CNNs can be followed by RBMH for compact and invariant hashes. (Communications of the ACM, ©2017 ACM, Inc. http://doi.acm.org/10.1145/3078971.3078987)

First, we present the proposed hashing pipeline which mainly consists of two parts: (a) an unsupervised dimensionality reduction approach using Restricted Boltzmann Machines (RBM) to produce binary hashes at the target bitrate (Sect. 3.1) and (b) a fine-tuning step to improve the binary embedding functions generated by stacked RBMs (Sect. 3.2). Specifically, the first dimensionality reduction step applies a regularization to RBM specifically designed to optimize the distribution of generated binary hash codes. The proposed approach is a batch-level regularization scheme aiming to improve very low bitrate hashes by encouraging efficient use of the latent subspace both within and across the hashes. The second fine-tuning step is based on metric refinement with Siamese networks, an idea originally proposed by Bromley et al. in [10]. The method is based on the use of a labeled training set of matching and nonmatching pairs of instances. Contrary to the pairwise contrastive loss function usually used at lower dimensionality such as in [17, 30], we show that critical improvements in the loss function of the Siamese network lead to improvements in search results.

In Sect. 3.3, we proposed Nested Invariance Pooling (NIP), a novel method based on *i-theory* for creating robust and compact global image descriptors from CNN for object instance search. The method provides a practical and mathematically proven way for computing invariant object representations with feed-forward neural networks. Through a thorough empirical study, we show that the incorporation of every new group invariance property following the method leads to consistent and significant improvements in search results. NIP has a few parameters (sequencing of transformations and choice of statistical moments are important), but experiments show that many default and reasonable settings produce results which can generalize well across all data sets, meaning that the risk of overfitting is low. This study also confirms the high potential of the feature pyramid ($pool5$) as a starting representation for high-performance compact hashes instead of the more commonly used first fully connected layer ($fc6$).

Finally, in Sect. 3.4, NIP is shown to be able to effectively combine with the RBM hashing scheme, leading to hashes that are both compact and robust to multiple types of image transformations. Thorough empirical evaluation with small- and large-scale datasets shows that the proposed scheme is able to produce extremely compact hashes that are able to outperform other schemes, especially at very low bitrates (32–256 bits).

3.1 Restricted Boltzmann Machine for Hashing

3.1.1 Method

Restricted Boltzmann Machine The Restricted Boltzmann Machine (RBM) [64] is a variant of the Boltzmann machine [34]. An RBM is an undirected bipartite graphical model consisting of a layer of visible units x and a layer of hidden or hidden units z. A set of symmetric weights W connects x and z. For an RBM with binary visible and hidden units, the joint set of visible and hidden units has an energy function given by:

$$E(x, z) = -\sum_i c_i x_i - \sum_j b_j z_j - \sum_{i,j} x_i z_j w_{ij} \tag{1}$$

where x_i and z_j are the binary states of visible and hidden units i and j, respectively, w_{ij} are the weights connecting the units, and c_i and b_j are their respective bias terms. Using the energy function in Eq. (1), a probability can be assigned to x as follows:

$$P(x) = \sum_z \frac{\exp(-E(x, z))}{Z}$$

where Z is a "partition" term, given by summing over all possible join sets of visible and hidden units:

$$Z = \sum_{x,z} \exp(-E(x, z))$$

The activation probabilities of units in one layer can be sampled by fixing the states of the other layer as follows:

$$P(z_j = 1|x) = f\left(b_j + \sum_i w_{ij} x_i\right) \tag{2}$$

Similarly, with symmetric weights:

$$P(x_i = 1|z) = f\left(c_i + \sum_j w_{ij} z_j\right) \tag{3}$$

where $f(\cdot)$ is the standard logistic function. RBMs can be trained by minimizing the contrastive divergence objective [32], which approximates the maximum likelihood of the input distribution. Alternating Gibbs sampling based on Eqs. (2) and (3) is used to obtain the network states to update the parameters w_{ij}, b_i, b_j through gradient descent.

Batch-Level Regularization Proper regularization is key during the training of RBM. For classification, discriminative performance is improved by constraining binary latent units to be rarely activated or sparse [55]. This is desirable for classification tasks as it improves separability of the data, but this is not necessarily ideal for hashing where efficient use of the limited latent subspace is key. The proposed RBMH method is a batch-level regularization scheme (unlike sparsity schemes which are usually instance level). It achieves efficient space use by controlling sparsity in a way to maximize the entropy not only within every hash but also between the same bit of different hashes. This effectively encourages:

1. Half the hash bits to be active for a given image
2. Each hash bit to be equiprobable across images

Correspondingly, we propose a batch-level regularization scheme, specifically designed for hashing, referred as RBMH. We first discuss how high batch-level entropy is encouraged in the RBMH framework. Next, we present how deep RBMH are constructed by stacking multiple RBMH. An overview diagram of the method is available in Fig. 4.

For two successive layers $l-1$ and l, and a batch B of input instances z_α^{l-1} with corresponding latent representations z_α^l, we define the regularization term adapted from the fine-grained regularization proposed in [22]:

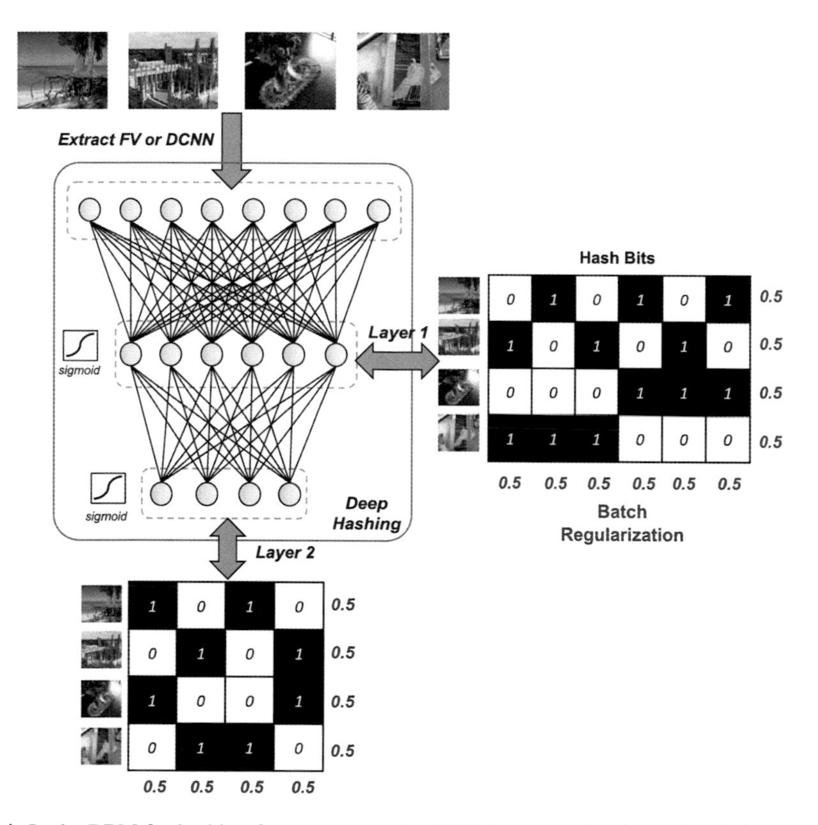

Fig. 4 In the RBM for hashing framework, stacked RBM are trained to hash global descriptors. The latent representations are regularized for the bits activation probability P to be equal to 0.5 both across bits of individual hashes and across images for the same latent unit. (©[2016] IEEE. Reprinted, with permission, from Ref.[46])

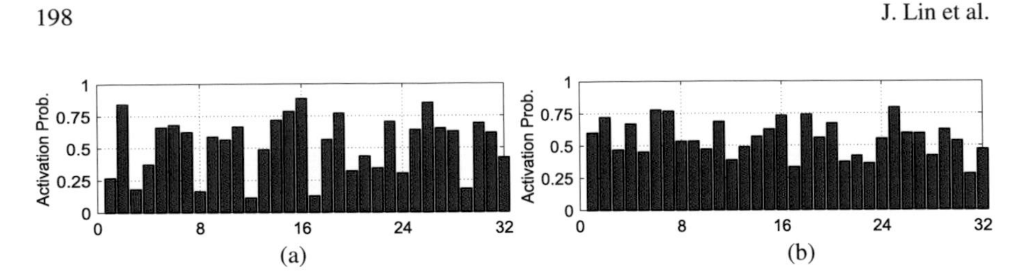

Fig. 5 Activation probabilities of hash bits with RBMH and the RBM proposed by Nair and Hinton [55]. Statistics of 32 bit binary hashes are computed on *Holidays* data set. The mean values of both activation histograms are close to 0.5, while the RBMH histogram is distributed more evenly across units compared to the standard RBM one (i.e., standard deviation is smaller). (**a**) Hinton RBM. (**b**) RBMH. (Communications of the ACM, ©2017 ACM, Inc. http://doi.acm.org/10.1145/3078971.3078987).

$$h(B) = -\sum_{z^l_\alpha \in B}\sum_j t^l_{j\alpha} \log z^l_{j\alpha} + (1 - t^l_{j\alpha}) \log(1 - z^l_{j\alpha}). \tag{4}$$

where $t^l_{j\alpha}$ are the target activations $z^l_{j\alpha}$ for each sample $z^{l-1}_{j\alpha}$. Unlike in [22], we choose the $t^l_{j\alpha}$ such that each $\{t^l_{j\alpha}\}_j$ for fixed α, and each $\{t^l_{j\alpha}\}_\alpha$ for fixed j is distributed according to $U(0, 1)$ effectively maximizing entropy. The uniform distribution is suitable for hashing high-dimensional vectors because the regularizer encourages each latent unit to be active with a mean of 0.5 while avoiding activation saturation.

The overall objective function for the RBMH becomes:

$$\underset{\{W^l, b^l, c^{l-1}\}}{\arg\min} \; -\sum_\alpha \log\Bigg(\sum_{z^l_\alpha \in E_\alpha} P(z^{l-1}_\alpha, z^l_\alpha) + \lambda h(B) \Bigg), \tag{5}$$

where λ is a regularization constant. It is optimized through batch gradient descent using the contrastive divergence algorithm.

Figure 5 shows the activation probabilities for 32-bit hashes provided by RBMH and the RBM proposed in [55]. The mean probability of activation is nearly 0.5 in both cases but much more evenly distributed across bits with RBMH.

Stacked RBMH The set of raw image representations lie in a complex manifold in a very high-dimensional feature space. Deeper networks have the potential to discover more complex nonlinear hash functions and improve object instance search performance. Following [33], we stack multiple RBMH, greedily training one layer at a time to create a deep network with several layers.

Each layer models the activation distribution of the previous layer and captures higher-order correlations between those units. For the hashing problem, we are interested in low-rate points of 64, 256, and 1024 bits. We progressively decrease the dimensionality of latent layers by a factor of 2^n per layer, where n is a tunable parameter. For our final models, n is empirically selected for each layer resulting in variable network depth.

Output Binarization Binary hashes are desirable for fast matching with Hamming distances. Therefore, sigmoid activation functions are used for the RBMH. In addition, the output of the topmost RBMH (layer L) is binarized around 0.5:

$$z_j^L = \begin{cases} 1, & \text{if } \frac{1}{1+\exp(-w_j^L z^{l-1} - b_j)} > 0.5 \\ 0, & \text{otherwise.} \end{cases} \quad (6)$$

In the next section, we evaluate the performance of the proposed SRBMH scheme.

3.1.2 Evaluation Framework

Global Descriptors SIFT [51] features obtained from Difference-of-Gaussian (DoG) interest points are used for FV. PCA is used to reduce dimensionality of the SIFT descriptor from 128 to 64 dimensions, which has shown to improve performance [38]. We use a Gaussian mixture model (GMM) with 128 centroids, resulting in 8192 dimensions each for first- and second-order statistics. Only the first-order statistics are retained in the global descriptor representation, as second-order statistics only results in a small improvement in performance [44]. The FV is L_2-normalized to unit-norm, after signed power normalization (referred to as FV from here on). DCNN features are extracted using the open-source software Caffe [40] with *AlexNet* reference model proposed for 2012 ImageNet classification task [41]. DCNN descriptors are extracted from the first fully connected layer $fc6$ which has been shown to yield performant descriptors for instance search. We refer to this 4096-dimensional descriptor as the CNN descriptor from here on.

SRBMH Training For the hashing problem, we are interested in low-rate points of 64, 256, and 1024 bits. SRBMH are trained greedily in a layer-by-layer fashion, i.e., each new RBMH is trained on the top of the previous one without modifying parameters of previous RBMH. A 150,000 images random subset of ImageNet is used as training data. The set is chosen for its variety and genericness and because it has no intersection with images used in the search experiments. The batch size is 100 for all experiments. Learning rate is set to 0.001 for the weight and bias parameters, momentum to 0.9. Training is run for a maximum of 30 epochs. For each rate point, different models are considered. For our final models, n is empirically selected for each layer resulting in variable network depth. Each target setting requires several hours to train.

Baselines Four state-of-the-art hashing schemes are considered as baselines: locality-sensitive hashing (LSH), iterative quantization (ITQ), bilinear projection binary codes (BPBC), and product quantization (PQ). LSH is based on random unit-norm projections of the raw descriptors, followed by signed binarization [73]. ITQ applies signed binarization after two transforms of raw descriptors: first the PCA, followed by a rotation [25]. Unlike ITQ, BPBC applies bilinear random projections, which require far less memory to transform the data [23].

For FV, we consider blocks of dimensions $D = 64, 256$ and 1024 and train $K = 256$ centroids for each block, resulting in $b = 128, 256$ and 64 bit descriptors, respectively. For CNN, we consider blocks of dimensions $D = 32, 128$ and 512, with $K = 256$ centroids, resulting in the same bitrates. L_2 norm is used for PQ and uncompressed descriptors, while Hamming distances are used for all binary hashing schemes.

Data Sets The *Holidays* and *UKBench* data sets are used for small-scale object instance search experiments. For large-scale experiments, the two data sets are combined with the one million *MIR FLICKR* distractor images.

Evaluation Metrics In most image search use cases, it is important for the relevant image to be present in the first step of the pipeline, matching global descriptors, so that a geometric consistency check (GCC) [19] step can subsequently find it. However, the GCC step is computationally complex and can only be performed on a small number of images. As a result, it is important for the relevant image to be present in a short list, so that the GCC step can find it. Hence, recall is presented at typical operating points of R = {10,100} and R = 1000 for small and large experiments, respectively. For *UKBench* small-scale experiments, 4 × Recall @ $R = 4$ is provided, to be consistent with the literature.

3.1.3 Experimental Results

Impact of Batch-Level Entropy Objective In Fig. 6a, we show the effect of applying the proposed regularization on a single-layer RBM 8192-b, for $b = 64, 256, 1024$. Hashing regularization significantly improves performance, ~10%

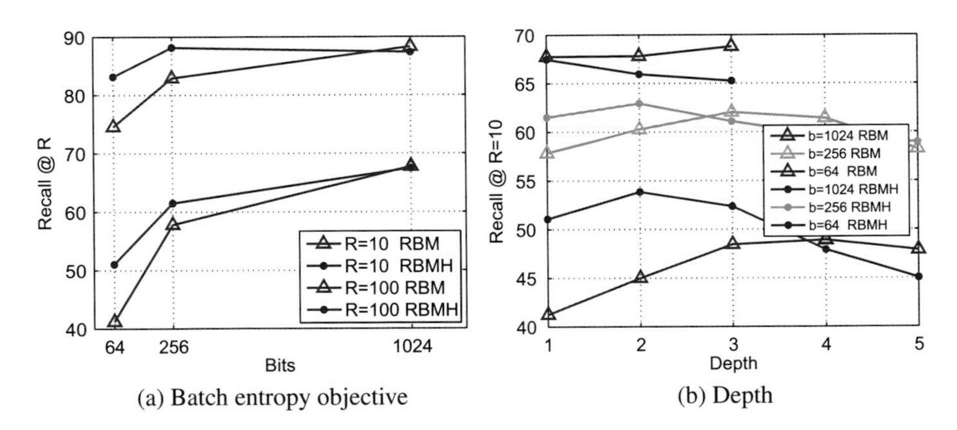

(a) Batch entropy objective (b) Depth

Fig. 6 Hashing *Holidays* FV. (**a**) RBMH regularization significantly improves performance for single-layer models 8192-b as b is decreased. (**b**) Recall improves as depth is increased for lower rate points $b = 64$ and $b = 256$. With RBMH regularization, same or better recall can be achieved at lower depth. (©[2016] IEEE. Reprinted, with permission, from Ref. [46])

absolute recall @ $R = 10$ at low-rate point $b = 64$. The performance gap increases as rate decreases. This is intuitive as the regularization encourages a more efficient use of the latent space.

Impact of Depth In Fig. 6b, we plot recall @ $R = 10$ on *Holidays* as depth is increased for a given rate point b. For $b = 1024$, we consider architectures $8192 - 1024$, $8192 - 4096 - 1024$, and $8192 - 4096 - 2048 - 1024$ corresponding to depth 1, 2, and 3, respectively. For rate points $b = 64$ and 256, similar configurations of varying depth are chosen. We observe that with no regularization, recall improves as depth is increased for $b = 256$ and $b = 64$, with optimal depth of three and four, respectively, beyond which performance drops. At higher rates of $b = 1024$ and beyond, increasing depth does not improve performance. For hashing, a sweet spot in performance for the depth parameter is observed for each rate point, as deeper networks can cause performance to drop due to loss of information over the layers. Similar trends are obtained for recall @ $R = 100$. Importantly, we observe that with the proposed regularization, we can achieve the same performance with lower depth at each rate point. This is critical, as the lower the depth, the faster the hash generation, and the lower the memory requirements.

Comparison with FV-RBMH and CNN-RBMH At a given rate point, CNN-RBMH outperforms FV-RBMH for all data sets, as shown in Fig. 7. At low rates, CNN-RBMH improves performance by more than 10% on the small data sets, possibly because CNN features are able to capture more complex low-level features and are a lower starting dimensionality compared to FV.

Comparison with Uncompressed Descriptors The performance of RBMH is compared to the uncompressed descriptor in Fig. 7. At 256 bits for CNN hashes, we only observe a marginal drop (a few percent) compared to the uncompressed descriptor for retrieval on all data sets. For FV, uncompressed descriptor performance is matched at 1024 bits. The instance search hashing problem becomes increasingly difficult as we move toward a 64-bit hash, with performance dropping steeply.

Comparison with State of the Art Small-scale search results are shown in Fig. 7. One can see that the proposed RBMH outperforms state of the art at most rates on all data sets, for both CNN and FV features. There is 2.4% improvement in absolute Recall @ $R = 100$ at $b = 64$ bits compared to the second performing scheme ITQ on *Holidays* for FV.

The performance ordering of other schemes depends on the bitrate and type of feature, while RBMH is consistent across data sets. Compared to ITQ scheme which applies a single PCA transform, each output bit for RBMH is generated by a series of projections. The PQ scheme performs poorly at the low rates in consideration, as large blocks of the global descriptor are quantized with a small number of centroids, as previously observed in [23]. LSH performs poorly at low rates but catches up given enough bits. Consistent trends are observed for the large-scale search in Fig. 8.

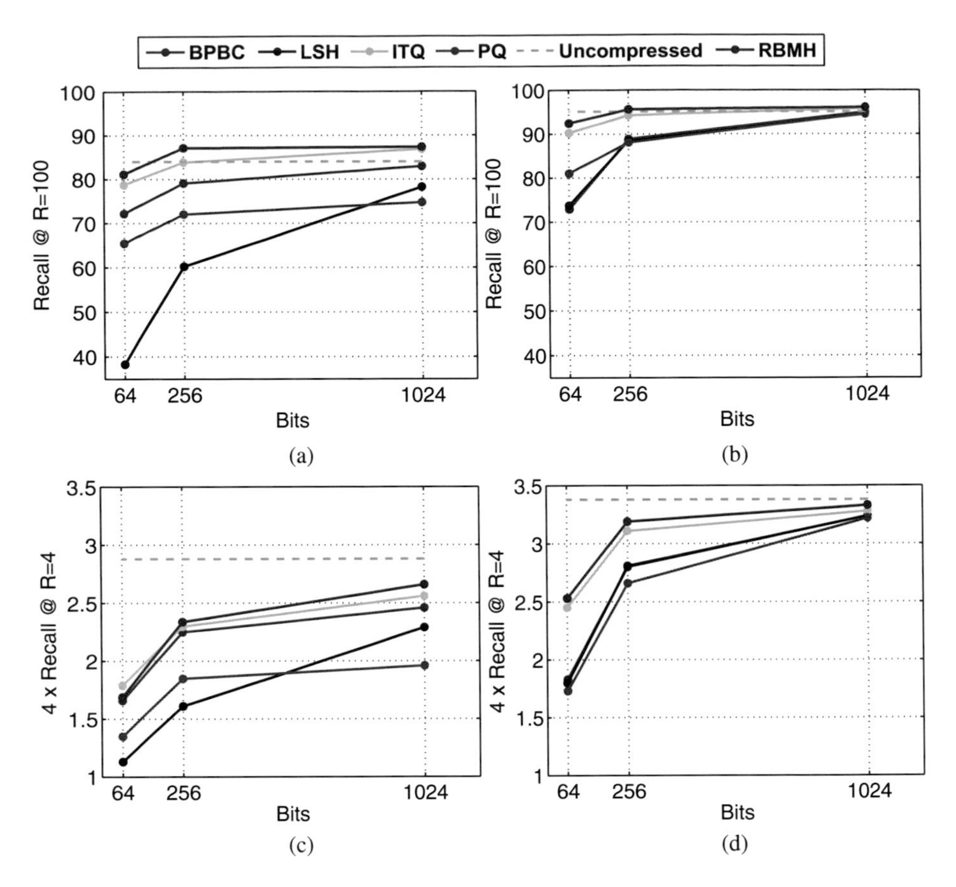

Fig. 7 Small-scale search results for different compression schemes. Proposed RBMH outperforms other schemes by a significant margin. (**a**) FV, *Holidays*. (**b**) CNN, *Holidays*. (**c**) FV, *UKBench*. (**d**) CNN, *UKBench*. (©[2016] IEEE. Reprinted, with permission, from Ref. [46])

3.2 Dual-Margin Siamese Fine-Tuning

In the previous section, we showed that global descriptors can be hashed into small binary descriptors using RBMH while retaining good properties for search. However, RBMH is specifically optimized for compression, and there is no built-in mechanism to ensure that the good metric properties of the original descriptors are preserved. In this section, we propose a weakly supervised method for improving the local structure of binary embedding functions using weight-sharing networks and an additional labeled data set of matching and nonmatching pairs (Fig. 9).

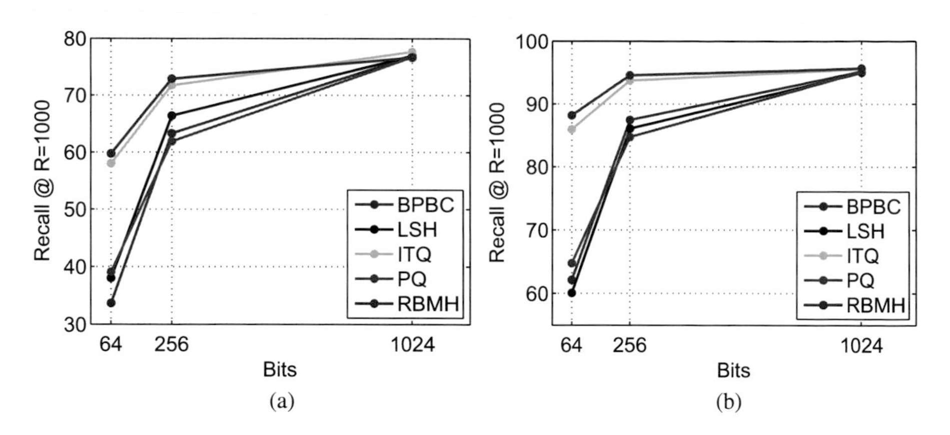

Fig. 8 Large-scale search results (with 1M distractor images, at Recall@R = 1000) for different compression schemes. *RBMH* outperforms other schemes at most rate points and data sets. (**a**) CNN, *Holidays* + 1M. (**b**) CNN, *UKBench* + 1M. (©[2016] IEEE. Reprinted, with permission, from Ref. [46])

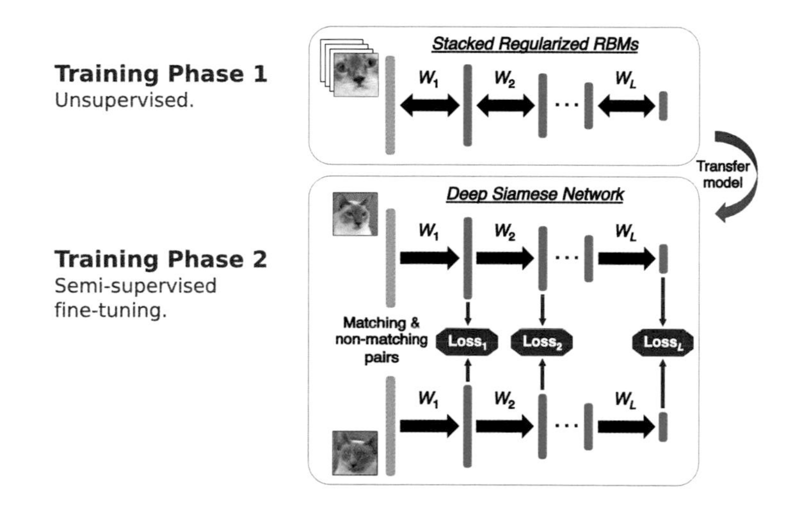

Fig. 9 The proposed method for learning binary embedding functions involves an unsupervised pre-training stage followed by a weakly supervised fine-tuning stage. In the first stage, RBM are trained in a layer-wise manner and stacked into a deep network (Sect. 3.1). In the second stage, first-stage parameters are loaded into a Siamese network and subsequently fine-tuned using matching and nonmatching pairs data. (Communications of the ACM, ©2017 ACM, Inc. http://doi.acm.org/10.1145/3078971.3078983)

3.2.1 Method

The fine-tuning is performed with a learning architecture known as Siamese networks first introduced in [30]. The principle was later successfully applied to deep architectures for face identification [17] and shown to produce representations robust to various transformations in the input space [30]. The use of Siamese architectures in the context of image search from CNN features was recently suggested as a possible improvement over the state of the art [9].

A Siamese network is a weakly supervised scheme for learning a similarity measure from pairs of data instances labeled as matching or nonmatching. In our adaptation of the concept, the weights of the trained RBM network are fine-tuned by learning a similarity measure at every intermediate layer in addition to the target space. Given a pair of data $(\mathbf{z}^0_\alpha, \mathbf{z}^0_\beta)$, a contrastive loss $\mathscr{D}_l$ is defined for every layer l, and the error is back propagated though gradient descent. Backpropagation for the losses of individual layers ($l = 1 \ldots L$) is performed at the same time. Applying the loss function proposed by Handsell et al. in [30] yields:

$$\mathscr{D}_l(\mathbf{z}^0_\alpha, \mathbf{z}^0_\beta) = y\|\mathbf{z}^l_\alpha - \mathbf{z}^l_\beta\|^2_2 + (1-y)\max(m - \|\mathbf{z}^l_\alpha - \mathbf{z}^l_\beta\|^2_2, 0) \tag{7}$$

where $y = 1$ if $(\mathbf{z}^0_\alpha, \mathbf{z}^0_\beta)$ is a matching pair or $y = 0$ otherwise, and $m > 0$ is a margin parameter affecting nonmatching pairs. As shown in Fig. 10a, the effect is to apply a contractive force between elements of any matching pairs and a repulsive force between elements of nonmatching pairs which element-wise distance is shorter than $\sqrt{m}$.

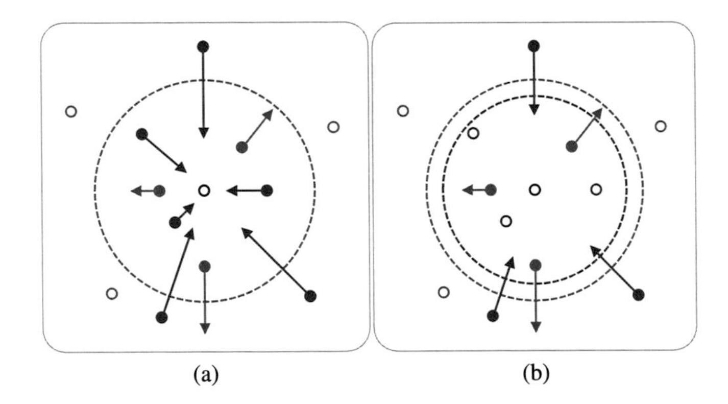

Fig. 10 A sample point (black dot) with corresponding matching (red dots) and nonmatching (blue dots) samples. The contrastive divergence loss used for fine-tuning can be interpreted as applying attractive forces between matching elements (red arrows) and repulsive forces between nonmatching elements (blue arrows). (**a**) The loss function (Eq. 7) proposed in [30] with a single margin parameter for nonmatching pairs (blue circle). Matching elements are subject to attractive forces regardless of whether they are already close enough from each other which adversely affects fine-tuning. (**b**) The proposed loss function (Eq. 8) with an additional margin parameter affecting matching pairs reciprocally (red circle). (Communications of the ACM, ©2017 ACM, Inc. http://doi.acm.org/10.1145/3078971.3078983)

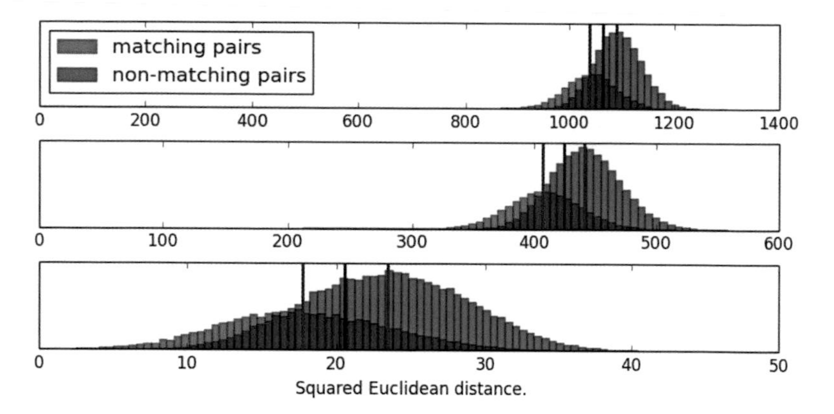

Fig. 11 Histograms of squared Euclidean distances for 20,000 matching pairs and corresponding 40,000 nonmatching pairs for an 8192 − 4096(top)−2048(middle)−64(bottom) stacked RBM network. Image pairs are sampled from *Yandex* data set. The red and blue vertical lines indicate the median values for the matching and nonmatching pairs, respectively. The Siamese loss shared margin value m is systematically set to be the mean of the two values (black vertical lines). (Communications of the ACM, ©2017 ACM, Inc. http://doi.acm.org/10.1145/3078971.3078983)

However, experiment results in Fig. 12 show that the loss function (7) causes a quick drop in search results. Results with nonmatching pairs alone suggest that the handling of matching pairs is responsible for the drop. The indefinite contraction of matching pairs well beyond what is necessary to distinguish them from nonmatching elements is a damaging behavior, especially in a fine-tuning context since the network is first globally optimized with a different objective. Figure 11 shows that any two elements, even matching, are always far apart in high dimension. Note that this phenomenon which occurs at the target bitrate of the hashes (e.g., 64 bits and higher) was not originally an issue at the much lower-dimensionality latent spaces considered in [30]

As a solution, we propose a double-margin loss with an additional parameter affecting matching pairs:

$$\begin{aligned}\mathscr{D}_l(\mathbf{z}^0_\alpha, \mathbf{z}^0_\beta) =& y \max(\|\mathbf{z}^l_\alpha - \mathbf{z}^l_\beta\|^2_2 - m_1, 0)\\ &+ (1-y)\max(m_2 - \|\mathbf{z}^l_\alpha - \mathbf{z}^l_\beta\|^2_2, 0)\end{aligned} \tag{8}$$

As shown in Fig. 10b, the new loss can thus be interpreted as learning "local large-margin classifiers" (if $m_1 \leq m_2$) to distinguish between matching and nonmatching elements. In practice, we found that the two margin parameters can be set equal ($m_1 = m_2 = m$) and tuned automatically from the statistical distribution of the sampled matching and nonmatching pairs (Figs. 11 and 12).

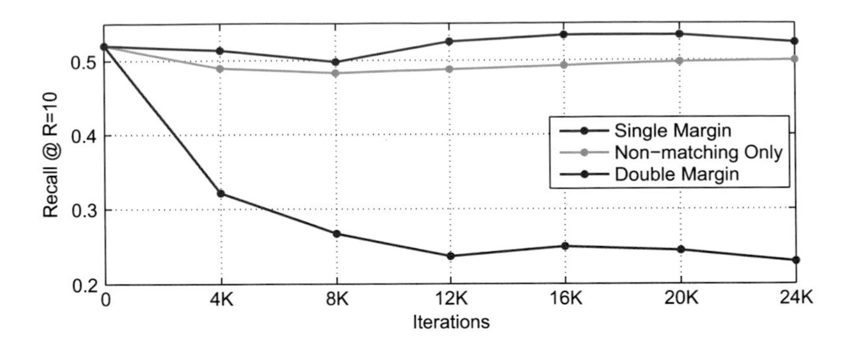

Fig. 12 Recall @ R = 10 on the *Holidays* data set. Over several iterations of Siamese fine-tuning. The recall rate quickly collapses when using the single margin loss function suggested in [30], while performance is better retained when only nonmatching pairs are passed. The double-margin loss solves the problem. The network is a stacked RBM (8192 – 4096 – 2048 – 64) trained with FV descriptors computed from a 150K random subset of ImageNet data set. Image pairs are sampled from the *Yandex* data set. For every matching pair, a random nonmatching element is chosen from the data set to form two nonmatching pairs. There are 33 matching pairs and 66 corresponding nonmatching pairs with every iteration. (Communications of the ACM, ©2017 ACM, Inc. http://doi.acm.org/10.1145/3078971.3078983)

3.2.2 Evaluation Framework

First, SRBM are trained as in Sect. 3.1, using the same FV descriptors extraction strategy. Then, a data set of matching and nonmatching image pairs is used for fine-tuning. The 200 K matching pair data set is provided by Yandex in their recent work [9]. It consists images of famous landmarks collected by querying the names of the most viewed Wikipedia landmark pages in Yandex search engine. Data set visualization reveals that most images depict buildings. For each matching pair, a random image is picked to generate two nonmatching pairs.

Holidays, *UKBench*, and *Oxbuild* data sets are used for small-scale search experiments. For large-scale experiments, *Holidays* and *UKBench* databases are combined with the one million *MIR FLICKR* distractor images. Note the *Yandex*-based training set is independent from the data used for evaluation as the authors removed Oxford-related queries and Holidays near-duplicate images.

3.2.3 Experimental Results

Detailed search results of a three-layer model before and after Siamese fine-tuning are provided in Table 1. The results show consistent improvements on all data sets and bitrates, with an average improvement of 2.78% (up to 6.24%). The difference is more significant at higher recall rates with an average of 2.43% @ R = 10 compared to 3.13% @ R = 100. They are however quite comparable when relative improvement rate is considered: 7.46% @ R = 10 and 7.24% @ R = 100 relatively.

Table 1 Small-scale search results before and after Siamese fine-tuning, with corresponding improvement. A SRBM model (8192 – 4096 – 2048 – 64) is trained on ImageNet FV descriptors and then fine-tuned on *Yandex*. For each model, the three layers search performance is evaluated on *Holidays*, *UKBench*, and *Oxbuild* data sets, before and after fine-tuning. Siamese fine-tuning consistently improves search performance at all three layers

		Recall @ R = 10			Recall @ R = 100		
Data set	Layer	bef.	aft.	diff.	bef.	aft.	diff.
Holidays	4096	70.83	73.67	**2.84**	89.92	91.40	**1.48**
	2048	67.74	71.12	**3.38**	88.77	92.04	**3.27**
	64	52.06	53.04	**0.98**	80.38	83.91	**3.53**
UKBench	4096	79.22	82.22	**3.00**	92.04	93.73	**1.69**
	2048	75.62	79.37	**3.75**	90.79	92.82	**2.03**
	64	47.94	49.25	**1.31**	73.02	73.94	**0.92**
Oxbuild	4096	19.38	21.73	**2.35**	41.09	45.19	**4.10**
	2048	14.32	17.23	**2.91**	36.03	41.03	**5.00**
	64	10.69	12.01	**1.32**	23.75	29.99	**6.24**

Table 2 Large-scale search results before and after Siamese fine-tuning, with corresponding differences. First, three SRBM are trained to, respectively, hash high-dimensional FV descriptors to bitrates $b = 1024$, 256, and 64. They are then fine-tuned with Siamese networks. Fine-tuning improves search performance at all bitrates on both data sets

		Recall @ R = 1000		
Data set	Bitrate	bef.	aft.	diff.
	1024	52.49	56.48	**3.99**
	256	46.07	49.60	**3.53**
Holidays + 1M	64	30.63	31.87	**1.24**
	1024	85.66	86.41	**0.75**
	256	78.70	80.74	**2.04**
UKBench + 1M	64	61.13	62.74	**1.61**

We notice differences across test sets with improvements on the *Oxford* set being more pronounced. The fine-tuning data set *Yandex*, like search data set *Oxbuild*, mostly depicts landmark structures. The proximity between the two data sets may explain the higher performance improvement on *Oxbuild*. The systematic improvements on all data sets are nevertheless evidence of the high transferability of both unsupervised training and semi-supervised fine-tuning.

Similar trends are observed with large-scale search experiments in Table 2, where Siamese fine-tuning improves performance at all bitrates and on all data sets.

3.3 Nested Invariance Pooling

In this section, we propose Nested Invariance Pooling (NIP), a method to produce compact global image descriptors from visual representations extracted from CNNs. The proposed method draws its inspiration from the *i-theory* [3–5], a mathematical theory for computing group-invariant transformations with feed-forward neural networks. The theory is an information processing model explaining how feed-forward information processing can be made robust to various types of signal distortions.

After showing that CNNs are compatible with the *i-theory*, we propose a simple and practical way to apply the theory to the construction of global image descriptors which are robust to various types of transformations of the input image at the same time. Through a thorough empirical evaluation based on multiple publicly available data sets, we show that proposed method is able to significantly consistently improve search results while keeping dimensionality low. Rotations, translations, and scale changes are studied in the scope of this section, but the proposed approach is extensible to other types of transformations. We show that using moments of increasing order for incorporating invariance to multiple transformation groups throughout nesting is important. Resulting NIP descriptors are invariant to various types of image transformations, and we show that the process significantly improves search results while keeping dimensionality low (512 dimensions).

3.3.1 I-Theory in a Nutshell

Many common classes of image transformations such as rotations, translations, and scale changes can be modeled by the action of a transformation group. Let an image $x \in E$ and a group G of transformations acting over E with group action $G \times E \rightarrow E$ denoted with a dot (.). The orbit of x by G is the subset of E defined as $O_x = \{g.x \in E | g \in G\}$. The orbit corresponds to the set of transformations of x under groups such as rotations, translations, and scale changes. It can be easily shown that O_x is globally invariant to the action of any element of G, and thus, any descriptor computed directly from O_x would be globally invariant to G.

The *i-theory* builds invariant representations for a given object $x \in E$ in relation with a predefined template $t \in E$ from the distribution of the dot products $D_{x,t} = \{< g.x, t > \in \mathbb{R} | g \in G\} = \{< x, g.t > \in \mathbb{R} | g \in G\}$ over the orbit. The following representation (for any $n \in \mathbb{N}^*$) is proven to have proper invariance and selectivity properties provided that the group is compact or locally compact:

$$\mu_{G,t,n}(x) = \frac{1}{\int_G dg} \left(\int_G | < g.x, t > |^n dg \right)^{\frac{1}{n}} \tag{9}$$

One may note that the transformation can be applied either on the image or the template indifferently. Note that the sequence $(\mu_{G,t,n}(x))_{n \in \mathbb{N}^*}$ is analogous to a histogram. Such a representation is mathematically proven to have proper invariance and selectivity properties provided that the group is compact or at least locally compact [4].

In practice, while a compact group (e.g., rotations) or locally compact group (e.g., translations, scale changes) is required for the theory to be mathematically provable, the authors of [4] suggest that the theory extends well (with approximate invariance) to nonlocally compact groups and even to continuous non-group transformations (e.g., out-of-plane rotations, elastic deformations) provided that proper class-specific templates can be provided. Recent work on face verification [43] and music classification [74] apply the theory to non-compact groups with good results.

3.3.2 CNNs Are I-Theory Compliant Networks

Popular CNN architectures designed for image classification such as *AlexNet* [41] and *OxfordNet* [63] share a common building block: a succession of convolution-pooling operations designed to model increasingly high-level visual representations of the data. The highest-level visual features may then be fed into fully connected layers acting as classifiers.

As shown in detail on Fig. 13a, the succession of convolution and pooling operations in a typical CNN is in fact a way to incorporate local translation invariance strictly compliant with the framework proposed by the *i-theory*. The network architecture provides the robustness such as predicted by the invariance theory, while training via backpropagation ensures a proper choice of templates. Multiple convolution-pooling steps are applied (five times in both *AlexNet* and *OxfordNet*) resulting in increased robustness and higher-level templates. Note that the iterative composition of local translation invariance approximately translates into robustness to local elastic distortions for the features at the *pool*5 layer.

In this study, instead of the popular first fully connected layer (*fc6*) which is on average the best single CNN layer to use as a global out-of-the-box descriptor for image search, we decided to use the locally invariant *pool*5 as a starting representation for the proposed global descriptors and further enhance their robustness to selected transformation groups in a way inspired from *i-theory*.

3.3.3 Multigroup-Invariant CNN Descriptors

We build the NIP descriptors starting from the already locally robust *pool5* feature maps of *OxfordNet*. Global invariance to several transformation groups is then sequentially incorporated following the *i-theory* framework. The specific transformation groups considered in this study are translations G_T, rotations G_R, and scale changes G_S. For every feature map i of the *pool5* layer ($0 \leq i < 512$), we denote $f_i(x)$ the corresponding unit's output. As shown on Fig. 13b, transformations g are applied on the input image x varying the output of the *pool5* feature $f_i(g.x)$. Note that the transformation f_i is nonlinear due to multiple convolution-pooling

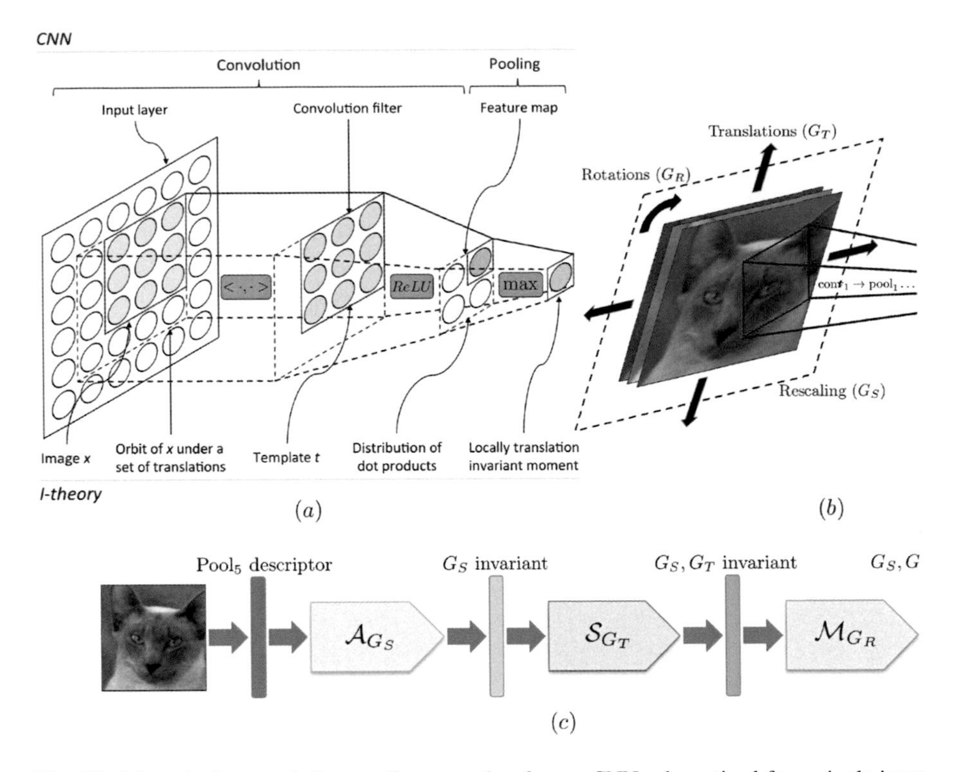

Fig. 13 (**a**) A single convolution-pooling operation from a CNN schematized for a single input layer and single output unit. The parallel with *i-theory* shows that the universal building block of CNN is compatible with the incorporation of invariance to local translations of the input according to the theory. The network architecture is responsible for the invariance properties, while backpropagation provides a practical way to learn the templates from the data. (**b**) A specific succession of convolution and pooling operations learnt by the CNN (depicted in red) computes the *pool*5 feature f_i for each feature map i from the RGB image data. A number of transformations g can be applied to the input x in order to vary the response $f_i(g.x)$. (**c**) The proposed method takes inspiration from the *i-theory* to create compact and robust global image descriptors from CNN. Starting with raw *pool*5 descriptors, it can be used to stack up an arbitrary number of transformation group invariances while keeping the dimensionality under control. The particular sequence of transformation groups and statistical moments represented on the diagram was shown to produce the best performing hashes on average, but other arbitrary combinations are also able to improve search results

operations; thus, it is not strictly a mathematical dot product but can still be viewed as an inner product. Accordingly, the pooling scheme used by NIP with $G \in \{G_T, G_R, G_S\}$ is:

$$\mathscr{X}_{G,i,n}(x) = \frac{1}{\int_G dg} \left(\int_G f_i(g.x)^n dg \right)^{\frac{1}{n}} \tag{10}$$

$$= \frac{1}{m} \left(\sum_{j=0}^{m-1} f_i(g_j.x)^n \right)^{\frac{1}{n}} \tag{11}$$

when O_x is discretized into m samples. The corresponding global image descriptors are obtained after each pooling step by concatenating the moments for the individual features:

$$\mathscr{X}_{G,n}(x) = (\mathscr{X}_{G,i,n}(x))_{0 \leq i < 512} \tag{12}$$

As shown in Eq. 11, the pooling operation has an order parameter n defining the "hardness" of the pooling. $n = 1$ is average pooling, while $n \to +\infty$ on the other extreme is max-pooling. $n = 2$ is analogous to standard deviation. Subsequently, we refer to the moments for $n = 1, 2, +\infty$ as $\mathscr{A}_G$, $\mathscr{S}_G$, and $\mathscr{M}_G$.

Work on *i-theory* [74] has shown that it is possible to chain multiple types of group invariances one after the other [74]. We apply this principle on NIP descriptors by making them invariant to several transformations. For instance, following scale invariance with average ($n = 1$) by translation invariance with hard max-pooling ($n \to +\infty$) is done by:

$$\max_{g_t \in G_T} \left(\frac{1}{\int_{g_s \in G_S} dg_s} \int_{g_s \in G_S} f_i(g_t g_s.x) dg_s \right) \tag{13}$$

$$= \max_{j \in [0, m_t - 1]} \left(\frac{1}{m_s} \sum_{i=0}^{m_s - 1} f_i(g_{t,j} g_{s,i} t.x) \right) \tag{14}$$

Operations are somctimes commutable (e.g., $\mathscr{A}_G$ and $\mathscr{A}_{G'}$) and sometimes not (e.g., $\mathscr{A}_G$ and $\mathscr{M}_{G'}$) depending on the specific combination of moments so the sequence of transformations does matter for NIP. The hardness parameter n must also be chosen carefully. Empirically, we found pooling progressively with increasing moments (e.g., $\mathscr{A}_G$, then $\mathscr{S}_G$, then $\mathscr{M}_G$) to work well, as presented in the experiments section.

Pairwise Matching Distance Object instance search starts with the construction of a list of database images ordered according to their pairwise matching distance with the query image. With CNN descriptors, the matching distance is strongly affected by commonly encountered image transformations. We observe that a rotation of the query image by more than 10 degrees causes a sharp drop in results. This particular issue is much less pronounced with the popular Fisher vectors, largely due to the use of interest point detectors.

Figure 14 provides an insight on how adding different types of invariance with the proposed method will affect the matching distance on different image pairs of matching objects. With the incorporation of each new transformation group, we notice that the relative reduction in matching distance is the most significant with the image pair, which is the most affected by the transformation group.

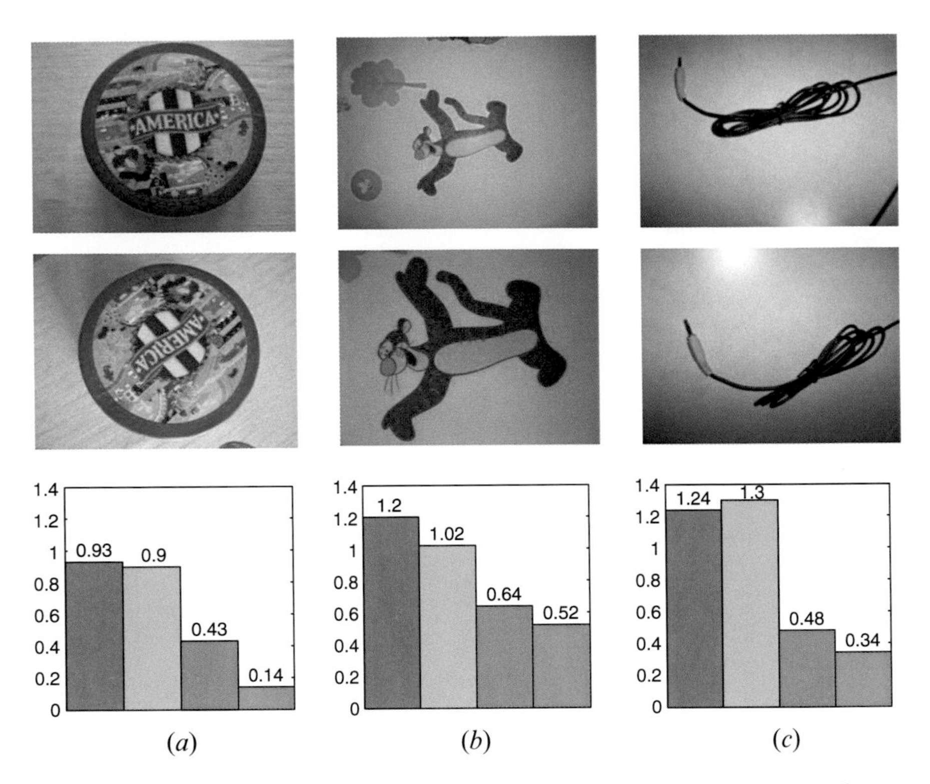

Fig. 14 Distances for three matching pairs from *UKBench*. For each pair, four pairwise distances (L_2-normalized) are computed corresponding to the following descriptors: *pool*5, $\mathscr{A}_{G_S}$, $\mathscr{A}_{G_S}$-$\mathscr{A}_{G_T}$, and $\mathscr{A}_{G_S}$-$\mathscr{A}_{G_T}$-$\mathscr{A}_{G_R}$. Adding scale invariance makes the most difference on (**b**), translation invariance on (**c**), and rotation on (**a**) which is consistent with the scenarios suggested by respective images pairs. (Communications of the ACM, ©2017 ACM, Inc. http://doi.acm.org/10.1145/3078971.3078987)

3.3.4 Evaluation Framework

The *pool*5 layer from the 16 layers *OxfordNet* [63] is chosen as starting representation, with a total dimensionality of 25,088 organized in 512 feature maps of size 7×7. For rotation invariance, rotated input images are padded with the mean pixel value computed from the *ImageNet* data set. The step size for rotations is 10 degrees yielding 36 rotated images per orbit. For scale changes, ten different center crops geometrically spanning from 100% to 50% of the total image have been taken. For translations, the entire feature map is used for every feature, resulting in an orbit size of $7 \times 7 = 49$.

We evaluate the instance search performance of the descriptors against four popular data sets: *Holidays*, *UKBench*, *Oxbuild*, and *Graphics*. The four data sets are chosen for the diversity of data they provide: *UKBench* and *Graphics* are object centric featuring close-up shots of objects in indoor environments. *Holidays* and *Oxbuild* are scene-centric data sets consisting primarily of outdoor buildings and scenes. Results are evaluated using mean average precision (mAP) and $4\times$ Recall @ $R = 4$ for *UKB*, to be consistent with the literature.

3.3.5 Experimental Results

Transformations, Order, and Moments As shown in Table 3, we first study the effects of incorporating various transformation groups and using different moments on descriptors. *Pool5* which is the starting point of NIP descriptors, and $fc6$ which is considered the best off-the-shelf descriptor [60] are provided as baselines. We present results for all possible combinations of transformation groups for average pooling (order does not matter as averages commute) and for the single best performer which is $\mathscr{A}_{G_S}$-$\mathscr{S}_{G_T}$-$\mathscr{M}_{G_R}$ (order matters).

First, we can immediately point out the high potential of $pool5$. Although it performs notably worse than $fc6$ as-is, a simple average pooling over the space of translations $\mathscr{A}_{G_T}$ makes it both better and eight times more compact than $fc6$. Similar observations have also been reported by [7, 8].

Second, as shown in Fig. 15, accuracy increases with the number of transformation groups involved. On average, single transformation schemes perform 21% better compared to $pool5$, 2-transformation schemes perform 34% better, and the 3-transformation scheme performs 41% better.

Table 3 Search results (mAP) for different sequences of transformation groups and moments. Results are computed with the mean average precision (mAP) metric. For reference, $4 \times$ Recall@4 results are also provided for *UKBench* (between parentheses). G_T, G_R, and G_S denote the groups of translations, rotations, and scale changes, respectively. Note that averages commute with other averages so the sequence order of the composition does not matter when only averages are involved. Best results are achieved by choosing specific moments. $\mathscr{A}$, $\mathscr{S}$, and $\mathscr{M}$ denote the moments average, standard deviation, and maximum, respectively. $\mathscr{A}_{G_S}$-$\mathscr{S}_{G_T}$-$\mathscr{M}_{G_R}$ corresponds to the best average performer. *fc6* and $pool5$ are provided as a baseline

		Data set			
Sequence	Dims	*Oxbuild*	*Holidays*	UKB	Graphics
pool5	25,088	0.427	0.707	0.823(3.11)	0.315
fc6	4096	0.461	0.782	0.910(3.50)	0.312
$\mathscr{A}_{G_T}$	512	0.477	0.800	0.924(3.56)	0.322
$\mathscr{A}_{G_R}$	25,088	0.462	0.779	0.954(3.72)	0.500
$\mathscr{A}_{G_S}$	25,088	0.430	0.716	0.828(3.12)	0.394
$\mathscr{A}_{G_T}$-$\mathscr{A}_{G_R}$	512	0.418	0.796	0.955(3.73)	0.417
$\mathscr{A}_{G_T}$-$\mathscr{A}_{G_S}$	512	0.537	0.811	0.931(3.61)	0.430
$\mathscr{A}_{G_R}$-$\mathscr{A}_{G_S}$	25,088	0.494	0.815	0.959(3.75)	0.552
$\mathscr{A}_{G_T}$-$\mathscr{A}_{G_R}$-$\mathscr{A}_{G_S}$	512	0.484	0.833	0.971(3.82)	0.509
$\mathscr{A}_{G_S}$-$\mathscr{S}_{G_T}$-$\mathscr{M}_{G_R}$	512	**0.592**	**0.838**	**0.975(3.84)**	**0.589**

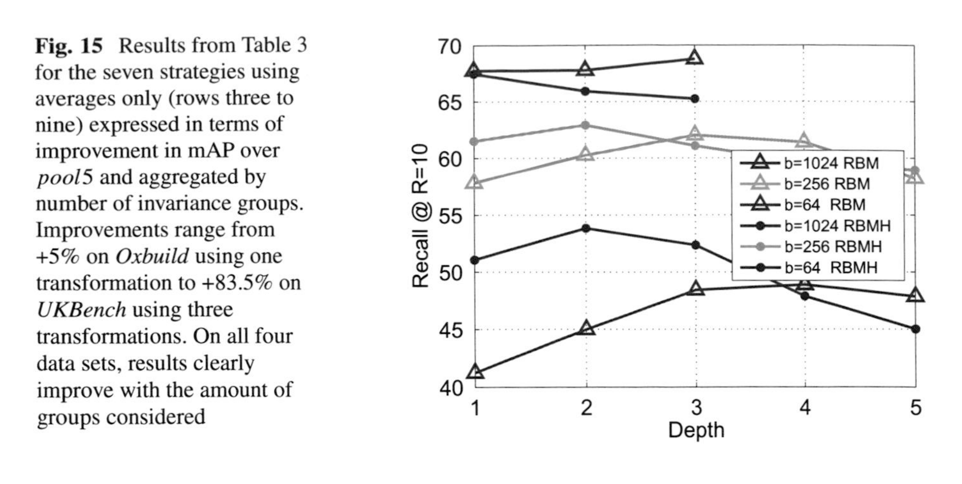

Fig. 15 Results from Table 3 for the seven strategies using averages only (rows three to nine) expressed in terms of improvement in mAP over *pool*5 and aggregated by number of invariance groups. Improvements range from +5% on *Oxbuild* using one transformation to +83.5% on *UKBench* using three transformations. On all four data sets, results clearly improve with the amount of groups considered

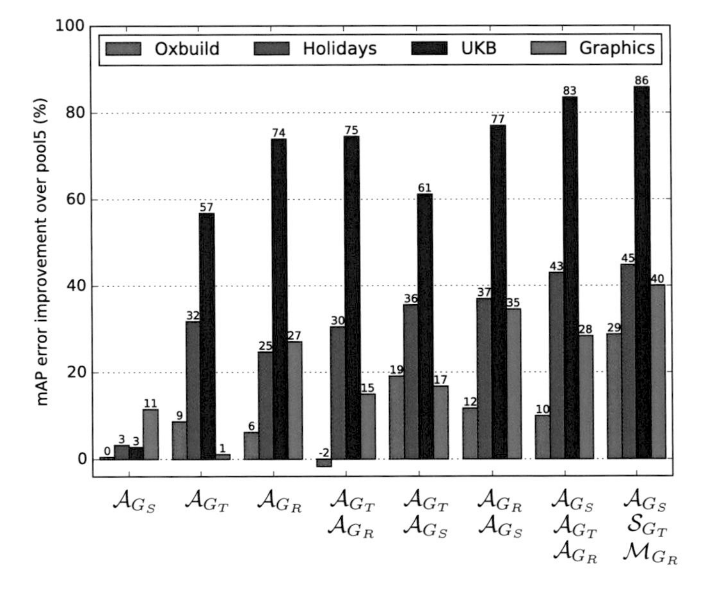

Fig. 16 Results from Table 3 expressed in terms of improvement in mAP over *pool*5. Most strategies yield significant improvements over *pool*5 on most data sets. The average improvement is 68% for the best strategy

Third, choosing statistical moments different than averages further improves the search results. In Fig. 16, we observe that $\mathscr{A}_{G_S}$-$\mathscr{S}_{G_T}$-$\mathscr{M}_{G_R}$ performs roughly 17% better (average results over all data sets) than $\mathscr{A}_{G_S}$-$\mathscr{A}_{G_T}$-$\mathscr{A}_{G_R}$. Notably, the best combination corresponds to an increase in the orders of the moments: $\mathscr{A}$ being a first-order moment, $\mathscr{S}$ second-order, and $\mathscr{M}$ of infinite order. A different way of stating this fact is that a more invariant representation requires a higher order of pooling. Overall, $\mathscr{A}_{G_S}$-$\mathscr{S}_{G_T}$-$\mathscr{M}_{G_R}$ improves results over starting representation *pool*5 by 39% (*Oxbuild*) to 87% (*Graphics*) depending on the data set. Better improvements with *Graphics* can be explained with the presence of many rotations in the data set (smaller objects taken under different angles), while *Oxbuild* consisting mainly of upright buildings is less significantly helped by incorporating rotation invariance.

Comparison with State of the Art State-of-the-art descriptors include variants of VLAD/FV [28, 39], deep descriptors [8, 9, 62, 66], and descriptors combining deep CNN and VLAD/FV [26, 57]. As shown in Table 4, we observe that 512-D NIP descriptors largely outperform most state-of-the-art methods with 512 or higher dimensions, on all data sets. Following [8, 62, 66], we also perform PCA whitening to reduce the dimensionality of NIP to 256. One can see that the 256-D NIP descriptors yield superior performance to [8, 62, 66] on all data sets.

First, we compare NIP to the most related papers [8, 62, 66] which propose 256-D deep descriptors by aggregating convolutional features with various pooling operations. With only one layer of pooling, [7, 8, 62] can be considered a special case of NIP, providing only limited levels of translation invariance. The recently proposed regional maximum activation of convolutions (R-MAC) [66] reports outstanding results on building data set *Oxbuild* with very small dimensionality (e.g., 0.668 mAP for 512-D R-MAC and 0.561 mAP for 256-D R-MAC). The authors propose a fast R-CNN-type pooling [20], which is effective when the object of interest is in a small portion of the image. Such an approach will be less effective when the object of interest is affected by groups of distortions like rotation and perspective and located at the center of the image. Here, we observe that nested pooling over many types of distortions with progressively increasing

Table 4 Search performance comparing NIP to other state-of-the-art methods. We include results in recent papers with comparable dimensionality of descriptors reported in those papers. L2 distance is used for all methods

Method	Dims	Data set		
		Oxbuild	*Holidays*	UKB
T-embedding [39]	1024	0.560	0.720	3.51
T-embedding [39]	512	0.528	0.700	3.49
FV + Proj [28]	512	–	0.789	3.36
FC + PCAWhitening [60]	500	0.322	0.642	–
FC + VLAD + PCA [26]	512	–	0.784	–
FC + Finetune + PCAWhitening [9]	512	0.557	0.789	3.30
Conv + MaxPooling [62]	256	0.533	0.716	–
FV + FC + PCAWhitening [57]	512	–	0.827	3.37
Conv + SPoC + PCAWhitening [8]	256	0.589	0.802	3.65
R-MAC + PCAWhitening [66]	512	**0.668**	–	–
R-MAC + PCAWhitening [66]	256	0.561	–	–
NIP	512	0.592	**0.838**	**3.84**
NIP + PCAWhitening	256	**0.609**	**0.836**	**3.83**

moments is essential to achieving geometric invariance and high search performance with low-dimensional descriptors. Besides, the technique proposed in [66] can be incorporated with NIP to further improve performance.

Next, we note that [62] reports better results on *Holidays* (0.881 mAP) and *Oxbuild* (0.844 mAP), with very high-dimensional descriptors (from 10K to 100K). These very high-dimensional descriptors are obtained by combining CNN descriptors with spatial max-pooling [7]. In contrast, NIP results are generated using only 256 to 512 dimensional descriptors.

3.4 Hashing with Invariant Descriptors

In this section, we combine NIP with RBMH (Fig. 3), the descriptor hashing scheme presented in the previous sections, with the aim to achieve the best performing tiny 32–256 bits hashes. Multigroup-invariant NIP representations, shown to be outstanding in previous section, are used as starting representations for hashing. RBMH are used then to produce compact binary hashes from the $\mathscr{A}_{G_S}$-$\mathscr{S}_{G_T}$-$\mathscr{M}_{G_R}$ 512-D floating point values descriptors.

Resulting hashes are compared to hashes obtained, using the same starting representation, with popular unsupervised hashing methods including ITQ [24], bilinear projection binary codes (BPBC) [23], PCAHash [24], LSH [18], SKLSH [59], SH [70], and RBM. Evaluated bitrates range from 32 to 256 bits. Original $\mathscr{A}_{G_S}$-$\mathscr{S}_{G_T}$-$\mathscr{M}_{G_R}$ 512-D floating point descriptors are also introduced as the baseline uncompressed scheme.

Experiments are conducted both with small-scale (Sect. 3.4) and large-scale (Sect. 3.4) search data sets. For small-scale object instance search experiments, four popular data sets are used: *Holidays*, *UKBench*, *Oxbuild*, and *Graphics*. For large-scale experiments, results are presented using the four data sets combined with the one million MIR FLICKR distractor images [35].

Finally, results of NIP + RBMH are also compared with other state-of-the-art methods in Sect. 3.4. NIP + RBMH are showed to be among the most compact and efficient binary codes for object instance search reported in the literature.

Small-Scale Experiments Small-scale search results of multigroup-invariant hashes are shown in Fig. 17, compared to other popular unsupervised hashing methods. RBMH codes outperform other methods at most code sizes on all data sets. First, there is a significant improvement at smaller code sizes like 32 bits, due to the proposed batch-level regularization: 0.457 vs. 0.369 in terms of mAP, compared to the second performing method RBM on *Holidays* at 32 bits. Second, the improvements of NIP + RBMH over other methods become smaller as code size increases (except SKLSH). For code size larger than 256 bits, the performances of all methods approach the upper bound, i.e., uncompressed descriptors. Finally, compared to uncompressed descriptors, there is a marginal drop for all methods on *UKBench* at 256 bits, while performance gap is larger for other data sets.

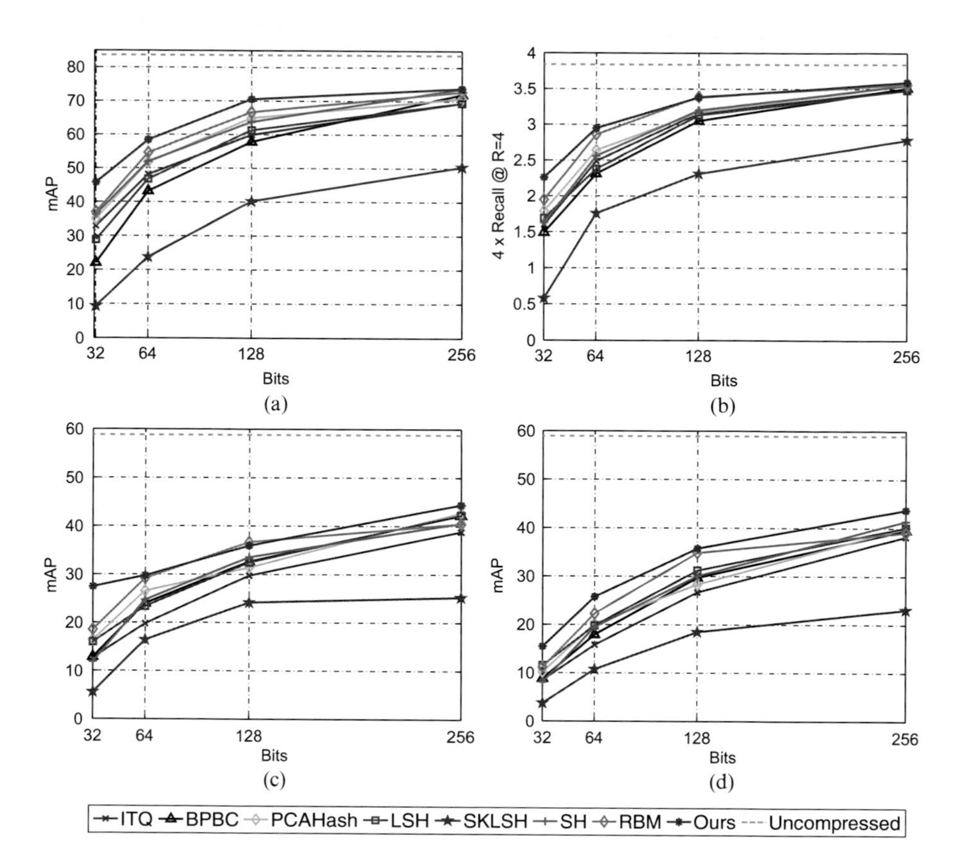

Fig. 17 Comparison of RBMH with other hashing methods on four benchmark data sets. All methods are built upon the best NIP descriptors. To examine the effect of compression, search results using uncompressed NIP descriptors are also presented. (**a**) *Holidays*. (**b**) UKB. (**c**) *Oxbuild*. (**d**) *Graphics*. (Communications of the ACM, ©2017 ACM, Inc. http://doi.acm.org/10.1145/3078971.3078987)

Large-Scale Experiments In Fig. 18, large-scale search results are presented, combining the one million MIR FLICKR distractor images with each data set, respectively. Trends consistent with small-scale search results in Fig. 17 are observed.

Comparison with State of the Art Invariant binary hashes are compared against different state-of-the-art pipelines, including methods compressing VLAD/FV with direct binarization [58], hashing [27], PQ [38, 76], and methods based on compact deep descriptors [9, 62].

As shown in Table 5, first, a simple binarization strategy (thresholding at data set mean) applied to $\mathscr{A}_{G_S}$-$\mathscr{S}_{G_T}$-$\mathscr{M}_{G_R}$ descriptor degrades search performance only very marginally and is sufficient to obtain significantly better accuracy than [9, 58]

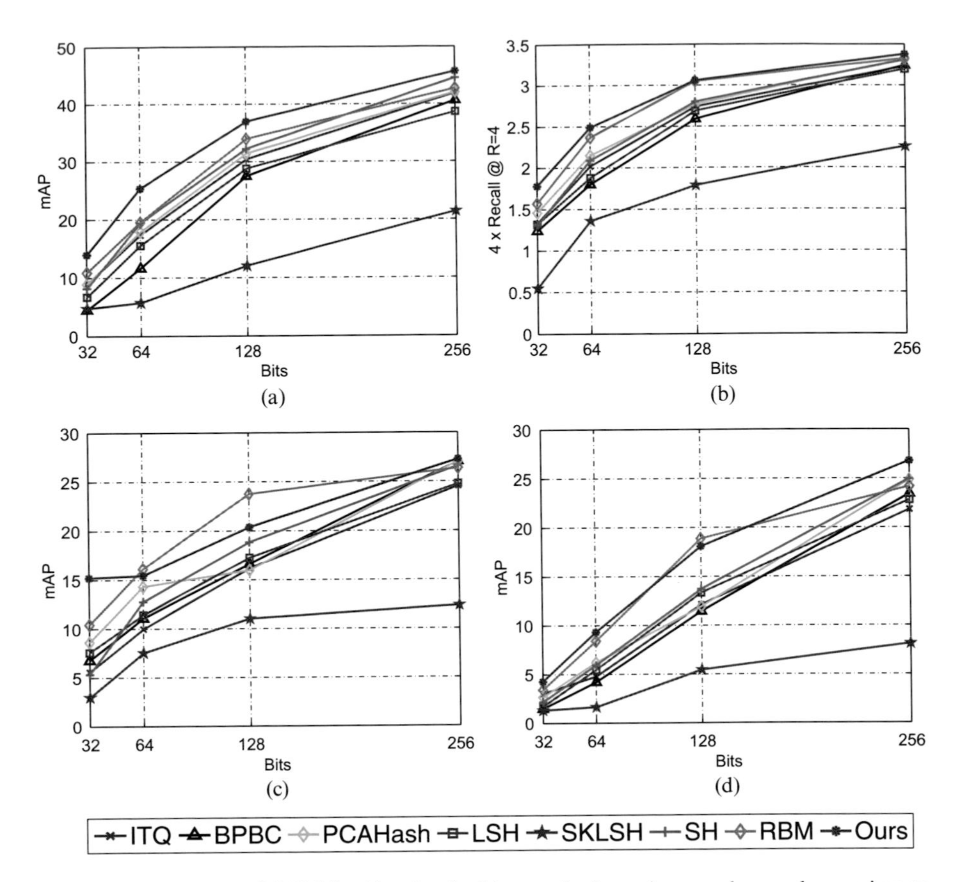

Fig. 18 Comparison of RBMH with other hashing methods on large-scale search experiments (with one million MIR FLICKR distractor images). All methods are based on the best NIP descriptors. (**a**) Holidays + 1M. (**b**) UKB + 1M. (**c**) Oxbuild + 1M. (**d**) Graphics + 1M. (Communications of the ACM, ©2017 ACM, Inc. http://doi.acm.org/10.1145/3078971.3078987)

at comparable code size (512 bits), e.g., 3.7 vs. 2.79 in [58] for 4× Recall @ $R = 4$ on *UKBench*.

Second, NIP + RBMH outperforms state of the art by a significant margin at comparable code sizes (from 32 to 256 bits). NIP+RBMH achieves the best performance on *Holidays* at small code size (128 bits), 0.705 vs. 0.644 mAP reported in the state of the art [75].

Note that Hamming distance is used for the multigroup-invariant binary descriptors, while other methods like PQ variants employ Euclidean distances (L2 or ADC), which typically result in higher accuracy than Hamming distance, at the expense of higher computational cost.

Table 5 Search performance comparing NIP+RBMH to other state-of-the-art methods at small code sizes (from 32 to 512 bits). ADC denotes asymmetric distance computation [27, 38]

Method	DIMS (size in bits)	Dist.	Data set		
			Oxbuild	*Holidays*	UKB
Binarized FV [58]	520(520)	Cosine	–	0.460	2.79
FV + SSH [27]	256(256)	ADC	–	0.544	3.08
FV + SSH [27]	128(128)	ADC	–	0.499	2.91
FV + SSH [27]	32(32)	ADC	–	0.334	2.18
FV + PQ [38]	128(128)	ADC	–	0.506	3.10
VLAD + PQ [75]	128(128)	L2	–	0.586	2.88
VLAD + CQ [75]	128(128)	L2	–	0.644	3.19
VLAD + SQ [76]	128(128)	L2	–	0.639	3.06
FC + Finetune + PCAWhitening [9]	16(512)	L2	0.418	0.609	2.41
Conv + MaxPooling [62]	256(256)	Cosine	0.436	0.578	–
Binarized NIP	512(512)	Hamming	**0.477**	**0.781**	**3.70**
NIP + RBMH	256(256)	Hamming	**0.445**	**0.739**	**3.59**
NIP + RBMH	128(128)	Hamming	**0.359**	**0.705**	**3.38**
NIP + RBMH	32(32)	Hamming	**0.274**	**0.458**	**2.26**

4 Conclusions and Future Works

In this chapter, we addressed the problem of CNN descriptors' limited robustness to geometric transformations, followed by dimensionality reduction and binarization of CNN descriptors. We first show how high-dimensional descriptors can be compressed to very compact binary representations in an unsupervised fashion using RBM and proposed a novel RBM regularization scheme for hashing (RBMH), key to achieving excellent performance at low rates. Next, we show how Siamese fine-tuning can be used to further improve hashes' performance and proposed a novel objective function suited for training Siamese networks in high-dimensional latent spaces settings. Subsequently, we proposed Nested Invariance Pooling (NIP), a novel method to produce global image descriptors from CNN. Finally, we show that NIP is compatible with the RBMH hashing scheme; combining the NIP+RBMH pipeline produces some of the best performing hashes available in the literature, especially at very low bitrates (256 bits and lower).

This thesis opens up several interesting avenues for future work.

- Pre-trained CNN models trained for large-scale image classification tasks, with larger amounts of data, have the potential of improving performance further for object instance search. For instance, CNN models trained on the full *ImageNet* data set with 14 million images, and 10,000 classes could lead to more discriminative features for the instance search task.
- While supervised CNN models have far outperformed their unsupervised CNN counterparts for large-scale image classification, the latter approach deserves careful attention in the context of instance search. For the instance search task, we

desire rich representations of low-level image information, which can be learnt directly from the large amounts of unlabeled image data available on the Internet. As image classification is not the end goal, unsupervised CNN models trained with large amounts of data might achieve comparable or better performance for instance search tasks. Availability of large amounts of training data (e.g., the Yahoo 100 million image data set [72]) and recent advances in open-source software for large-scale distributed deep learning (e.g., TensorFlow [2]) will enable training of large-scale unsupervised CNN models. If unsupervised CNN models work well for instance search, they will enable easier training and adaptation to different types of image databases.
- Rotation and scale invariance are key to instance search tasks. While the pooling schemes proposed in this work are highly effective, they are more of an afterthought to solving the invariance problem in the CNN context. Learning CNN representations which are inherently scale and rotation invariant is an exciting direction to pursue.
- Interest point detectors provide an efficient and effective way of achieving desired levels of invariance (ranging from scale and rotation invariance to affine invariance). The carefully handcrafted SIFT descriptor has been remarkably effective for the instance search task; however, patch-level descriptors can now be learnt with large amounts of data, using data sets like the Winder and Brown patch data sets [71] and the Stanford Mobile Visual Search patch data set [1]. A hybrid approach of interest point detectors with learnt CNN descriptor representations could lead to a significant improvement in search performance.
- Hybrid interest point detection schemes like the dense interest point detector proposed originally in [68] need to be revisited, in light of the effectiveness of CNN features which are extracted by dense sampling in the image. A recent survey of dense interest point detectors [68] is a good starting point.

References

1. CDVS Patches (2013). http://blackhole1.stanford.edu/vijayc/cdvs_patches.tar
2. Abadi, M., Agarwal, A., Barham, P., Brevdo, E., Chen, Z., Citro, C., Corrado, G.S., Davis, A., Dean, J., Devin, M., et al.: Tensorflow: Large-scale machine learning on heterogeneous distributed systems. arXiv preprint arXiv:1603.04467 (2016)
3. Anselmi, F., Leibo, J.Z., Rosasco, L., Mutch, J., Tacchetti, A., Poggio, T.: Magic materials: a theory of deep hierarchical architectures for learning sensory representations. CBCL paper (2013)
4. Anselmi, F., Leibo, J.Z., Rosasco, L., Mutch, J., Tacchetti, A., Poggio, T.: Unsupervised learning of invariant representations in hierarchical architectures. arXiv preprint arXiv:1311.4158 (2013)
5. Anselmi, F., Poggio, T.: Representation learning in sensory cortex: a theory. Tech. rep., Center for Brains, Minds and Machines (CBMM) (2014)
6. Arandjelovic, R., Zisserman, A.: All about vlad. In: Computer Vision and Pattern Recognition (CVPR), pp. 1578–1585 (2013)
7. Azizpour, H., Razavian, A., Sullivan, J., Maki, A., Carlsson, S.: From generic to specific deep representations for visual recognition. In: Computer Vision and Pattern Recognition Workshops (CVPR), pp. 36–45 (2015)

8. Babenko, A., Lempitsky, V.: Aggregating local deep features for image retrieval. In: International Conference on Computer Vision (ICCV), pp. 1269–1277 (2015)
9. Babenko, A., Slesarev, A., Chigorin, A., Lempitsky, V.: Neural codes for image retrieval. In: European Conference on Computer Vision (ECCV), pp. 584–599. Springer (2014)
10. Bromley, J., Guyon, I., LeCun, Y., Sackinger, E., Shah, R.: Signature verification using a Siamese time delay neural network. In: J. Cowan, G. Tesauro (eds.) Neural Information Processing Systems (NIPS), vol. 6. Morgan Kaufmann (1993)
11. Chandrasekhar, V., Lin, J., Morère, O., Goh, H., Veillard, A.: A practical guide to CNNs and Fisher vectors for image instance retrieval. Signal Processing (SIGPRO) (2016)
12. Chandrasekhar, V., Lin, J., Morère, O., Veillard, A., Goh, H.: Compact global descriptors for visual search. In: Data Compression Conference (DCC), pp. 333–342. IEEE (2015)
13. Chandrasekhar, V., Makar, M., Takacs, G., Chen, D., Tsai, S.S., Cheung, N.M., Grzeszczuk, R., Reznik, Y., Girod, B.: Survey of sift compression schemes. In: Mobile Multimedia Processing Workshop (MMP), pp. 35–40. Citeseer (2010)
14. Chandrasekhar, V., Takacs, G., Chen, D., Tsai, S., Grzeszczuk, R., Girod, B.: Chog: Compressed histogram of gradients a low bit-rate feature descriptor. In: Computer Vision and Pattern Recognition (CVPR), pp. 2504–2511. IEEE (2009)
15. Chen, D., Tsai, S., Chandrasekhar, V., Takacs, G., Chen, H., Vedantham, R., Grzeszczuk, R., Girod, B.: Residual enhanced visual vectors for on-device image matching. Conference Record of the Forty Fifth Asilomar Conference on Signals, Systems and Computers (ASILOMAR) pp. 850–854 (2011). URL http://ieeexplore.ieee.org/lpdocs/epic03/wrapper.htm?arnumber=6190128
16. Chen, D., Tsai, S., Chandrasekhar, V., Takacs, G., Vedantham, R., Grzeszczuk, R., Girod, B.: Residual enhanced visual vector as a compact signature for mobile visual search. Signal Processing (SIGPRO) **93**(8), 2316–2327 (2013)
17. Chopra, S., Hadsell, R., LeCun, Y.: Learning a similarity metric discriminatively, with application to face verification. In: Computer Vision and Pattern Recognition (CVPR), vol. 1, pp. 539–546. IEEE (2005)
18. Datar, M., Immorlica, N., Indyk, P., Mirrokni, V.S.: Locality-sensitive hashing scheme based on p-stable distributions. In: Annual Symposium on Computational Geometry (SoCG), pp. 253–262. ACM (2004)
19. Fischler, M.A., Bolles, R.C.: Random sample consensus: a paradigm for model fitting with applications to image analysis and automated cartography. Communications of the ACM **24**(6), 381–395 (1981)
20. Girshick, R.: Fast r-cnn. In: International Conference on Computer Vision (ICCV), pp. 1440–1448 (2015)
21. Girshick, R., Donahue, J., Darrell, T., Malik, J.: Rich feature hierarchies for accurate object detection and semantic segmentation. In: Computer Vision and Pattern Recognition (CVPR), pp. 580–587. IEEE (2014)
22. Goh, H., Thome, N., Cord, M., Lim, J.H.: Unsupervised and supervised visual codes with restricted Boltzmann machines. In: European Conference on Computer Vision (ECCV), pp. 298–311. Springer (2012)
23. Gong, Y., Kumar, S., Rowley, H., Lazebnik, S.: Learning binary codes for high-dimensional data using bilinear projections. In: Computer Vision and Pattern Recognition (CVPR), pp. 484–491 (2013)
24. Gong, Y., Lazebnik, S.: Iterative quantization: A procrustean approach to learning binary codes. In: Computer Vision and Pattern Recognition (CVPR), pp. 817–824. IEEE (2011)
25. Gong, Y., Lazebnik, S.: Iterative quantization: A procrustean approach to learning binary codes. Computer Vision and Pattern Recognition (CVPR) pp. 817–824 (2011)
26. Gong, Y., Wang, L., Guo, R., Lazebnik, S.: Multi-scale orderless pooling of deep convolutional activation features. In: European Conference on Computer Vision (ECCV), pp. 392–407. Springer (2014)
27. Gordo, A., Perronnin, F., Gong, Y., Lazebnik, S.: Asymmetric distances for binary embeddings. Pattern Analysis and Machine Intelligence (PAMI) **36**(1), 33–47 (2014)

28. Gordo, A., Rodríguez-Serrano, J.A., Perronnin, F., Valveny, E.: Leveraging category-level labels for instance-level image retrieval. In: Computer Vision and Pattern Recognition (CVPR), pp. 3045–3052. IEEE (2012)
29. Grauman, K., Fergus, R.: Learning binary hash codes for large-scale image search. In: Machine Learning for Computer Vision, pp. 49–87. Springer (2013)
30. Hadsell, R., Chopra, S., LeCun, Y.: Dimensionality reduction by learning an invariant mapping. In: Computer Vision and Pattern Recognition (CVPR), vol. 2, pp. 1735–1742. IEEE (2006)
31. Heo, J.P., Lee, Y., He, J., Chang, S.F., Yoon, S.E.: Spherical hashing. In: Computer Vision and Pattern Recognition (CVPR), pp. 2957–2964. IEEE (2012)
32. Hinton, G.E.: Training products of experts by minimizing contrastive divergence. Neural Computation (NC) **14**(8), 1771–1800 (2002)
33. Hinton, G.E., Osindero, S., Teh, Y.W.: A fast learning algorithm for deep belief nets. Neural Computation (NC) **18**(7), 1527–1554 (2006)
34. Hinton, G.E., Sejnowski, T.J.: Learning and relearning in Boltzmann machines. Parallel distributed processing: Explorations in the microstructure of cognition **1**, 282–317 (1986)
35. Huiskes, M.J., Thomee, B., Lew, M.S.: New trends and ideas in visual concept detection: the mir flickr retrieval evaluation initiative. In: Multimedia Information Retrieval (MIR), pp. 527–536. ACM (2010)
36. Jégou, H., Chum, O.: Negative evidences and co-occurences in image retrieval: The benefit of pca and whitening. In: European Conference on Computer Vision (ECCV), pp. 774–787. Springer (2012)
37. Jégou, H., Douze, M., Schmid, C.: On the burstiness of visual elements. In: Computer Vision and Pattern Recognition (CVPR), pp. 1169–1176. IEEE (2009)
38. Jégou, H., Perronnin, F., Douze, M., Sanchez, J., Perez, P., Schmid, C.: Aggregating local image descriptors into compact codes. Pattern Analysis and Machine Intelligence (PAMI) **34**(9), 1704–1716 (2012)
39. Jégou, H., Zisserman, A.: Triangulation embedding and democratic aggregation for image search. In: Computer Vision and Pattern Recognition (CVPR), pp. 3310–3317. IEEE (2014)
40. Jia, Y., Shelhamer, E., Donahue, J., Karayev, S., Long, J., Girshick, R., Guadarrama, S., Darrell, T.: Caffe: Convolutional architecture for fast feature embedding. In: Conference on Multimedia (ACMMM), pp. 675–678. ACM (2014)
41. Krizhevsky, A., Sutskever, I., Hinton, G.E.: Imagenet classification with deep convolutional neural networks. Neural Information Processing Systems (NIPS) pp. 1–9 (2012). URL https://papers.nips.cc/paper/4824-imagenet-classification-with-deep-convolutional-neural-networks.pdf
42. Kulis, B., Grauman, K.: Kernelized locality-sensitive hashing for scalable image search. In: International Conference on Computer Vision (ICCV), pp. 2130–2137. IEEE (2009)
43. Liao, Q., Leibo, J.Z., Poggio, T.: Learning invariant representations and applications to face verification. In: Neural Information Processing Systems (NIPS), pp. 3057–3065 (2013)
44. Lin, J., Duan, L.Y., Huang, T., Gao, W.: Robust Fisher codes for large scale image retrieval. In: International Conference on Acoustics and Signal Processing (ICASSP) (2013)
45. Lin, J., Duan, L.y., Wang, S., Bai, Y., Lou, Y., Chandrasekhar, V., Huang, T., Kot, A., Gao, W.: Hnip: Compact deep invariant representations for video matching, localization and retrieval. Transactions on Multimedia (TMM) (2017)
46. Lin, J., Morère, O., Chandrasekhar, V., Veillard, A., Goh, H.: Co-sparsity regularized deep hashing for image instance retrieval. In: International Conference on Image Processing (ICIP). IEEE (2016)
47. Lin, J., Morère, O., Petta, J., Chandrasekhar, V., Veillard, A.: Tiny descriptors for image retrieval with unsupervised triplet hashing. In: Data Compression Conference (DCC) (2016)
48. Lin, J., Morère, O., Veillard, A., Duan, L.Y., Goh, H., Chandrasekhar, V.: Deephash for image instance retrieval: Getting regularization, depth and fine-tuning right. In: ACM International Conference on Multimedia Retrieval (2017)
49. Liu, W., Wang, J., Ji, R., Jiang, Y.G., Chang, S.F.: Supervised hashing with kernels. In: Computer Vision and Pattern Recognition (CVPR), pp. 2074–2081. IEEE (2012)

50. Lou, Y., Bai, Y., Lin, J., Wang, S., Chen, J., Chandrasekhar, V., Duan, L.y., Tiejun, H., Kot, A., Gao, W.: Compact deep invariant descriptors for video retrieval. In: Data Compression Conference (DCC) (2017)
51. Lowe, D.: Distinctive image features from scale-invariant keypoints. International Journal of Computer Vision (IJCV) **60**(2), 91–110 (2004)
52. Lowe, D.G.: Distinctive image features from scale-invariant keypoints. International Journal of Computer Vision (IJCV) **60**(2), 91–110 (2004). doi:10.1023/B:VISI.0000029664.99615.94. URL http://link.springer.com/10.1023/B:VISI.0000029664.99615.94
53. Morère, O., Lin, J., Veillard, A., Duan, L.y., Chandrasekhar, V., Poggio, T.: Nested invariance pooling and rbm hashing for image instance retrieval. In: ACM International Conference on Multimedia Retrieval (2017)
54. Morère, O., Veillard, A., Lin, J., Petta, J., Chandrasekhar, V., Poggio, T.: Group invariant deep representations for image instance retrieval. In: AAAI Symposium on Science of Intelligence (2017)
55. Nair, V., Hinton, G.E.: 3D object recognition with deep belief nets. In: Neural Information Processing Systems (NIPS), pp. 1339–1347 (2009)
56. Norouzi, M., Blei, D.M.: Minimal loss hashing for compact binary codes. In: International Conference on Machine Learning (ICML), pp. 353–360 (2011)
57. Perronnin, F., Larlus, D.: Fisher vectors meet neural networks: A hybrid classification architecture. In: Computer Vision and Pattern Recognition (CVPR), pp. 3743–3752 (2015)
58. Perronnin, F., Liu, Y., Sánchez, J., Poirier, H.: Large-scale image retrieval with compressed fisher vectors. In: Computer Vision and Pattern Recognition (CVPR), pp. 3384–3391. IEEE (2010)
59. Raginsky, M., Lazebnik, S.: Locality-sensitive binary codes from shift-invariant kernels. In: Neural Information Processing Systems (NIPS), pp. 1509–1517 (2009)
60. Razavian, A., Azizpour, H., Sullivan, J., Carlsson, S.: CNN features off-the-shelf: an astounding baseline for recognition. In: Computer Vision and Pattern Recognition Workshop (CVPR), pp. 806–813 (2014)
61. Salakhutdinov, R., Mnih, A., Hinton, G.: Restricted Boltzmann machines for collaborative filtering. In: International Conference on Machine Learning (ICML), pp. 791–798. ACM (2007)
62. Sharif Razavian, A., Sullivan, J., Maki, A., Carlsson, S.: A baseline for visual instance retrieval with deep convolutional networks. In: International Conference on Learning Representations (ICLR). ICLR (2015)
63. Simonyan, K., Zisserman, A.: Very deep convolutional networks for large-scale image recognition. In: International Conference on Learning Representations (ICLR) (2015)
64. Smolensky, P.: Information processing in dynamical systems: foundations of harmony theory. In: Parallel distributed processing: explorations in the microstructure of cognition, pp. 194–281. MIT Press (1986)
65. Tolias, G., Avrithis, Y., Jégou, H.: To aggregate or not to aggregate: Selective match kernels for image search. In: International Conference on Computer Vision (ICCV), pp. 1401–1408 (2013)
66. Tolias, G., Sicre, R., Jégou, H.: Particular object retrieval with integral max-pooling of CNN activations. In: International Conference on Learning Representations (ICLR) (2016)
67. Torralba, A., Fergus, R., Weiss, Y.: Small codes and large image databases for recognition. In: Computer Vision and Pattern Recognition (CVPR), pp. 1–8. IEEE (2008)
68. Tuytelaars, T.: Dense interest points. In: Computer Vision and Pattern Recognition (CVPR), pp. 2281–2288. IEEE (2010)
69. Wang, J., Kumar, S., Chang, S.F.: Semi-supervised hashing for scalable image retrieval. In: Computer Vision and Pattern Recognition (CVPR), pp. 3424–3431. IEEE (2010)
70. Weiss, Y., Torralba, A., Fergus, R.: Spectral hashing. In: Neural Information Processing Systems (NIPS), pp. 1753–1760 (2009)
71. Winder, S., Hua, G., Brown, M.: Picking the best daisy. In: Computer Vision and Pattern Recognition (CVPR), pp. 178–185. IEEE (2009)

72. Yahoo: Yahoo! 100 million image data set. http://webscope.sandbox.yahoo.com/
73. Yeo, C., Ahammad, P., Ramchandran, K.: Rate-efficient visual correspondences using random projections. In: International Conference on Image Processing (ICIP), pp. 217–220. IEEE (2008)
74. Zhang, C., Evangelopoulos, G., Voinea, S., Rosasco, L., Poggio, T.: A deep representation for invariance and music classification. In: Acoustics, Speech and Signal Processing (ICASSP), pp. 6984–6988. IEEE (2014)
75. Zhang, T., Du, C., Wang, J.: Composite quantization for approximate nearest neighbor search. In: International Conference on Machine Learning (ICML), pp. 838–846 (2014)
76. Zhang, T., Qi, G.J., Tang, J., Wang, J.: Sparse composite quantization. In: Computer Vision and Pattern Recognition (CVPR), pp. 4548–4556 (2015)